William E. Gibbs

Lighting by Acetylene

Generators, Burners, and Electric Furnaces. Second Edition

William E. Gibbs

Lighting by Acetylene
Generators, Burners, and Electric Furnaces. Second Edition

ISBN/EAN: 9783337249472

Printed in Europe, USA, Canada, Australia, Japan

Cover: Foto ©berggeist007 / pixelio.de

More available books at **www.hansebooks.com**

Lighting by Acetylene

GENERATORS, BURNERS AND ELECTRIC FURNACES

BY

WILLIAM E. GIBBS, M.E.

SECOND EDITION, REVISED AND ENLARGED

NEW YORK
D. VAN NOSTRAND COMPANY

LONDON
CROSBY LOCKWOOD AND SON
1899

TABLE OF CONTENTS

INTRODUCTION TO SECOND EDITION

WHEN the first edition of this treatise was written, in the months of January and February, 1898, a host of acetylene generators was appearing. New types and modifications of older forms were being put upon the market, while hundreds were rushing into the new field opened by the production of calcium carbide. The apparent simplicity of the gas-making process from this new substance was sufficiently attractive to excite the interest of inventors generally, and to develop a large number of worthless generators, constructed by those having little mechanical knowledge and no ability to solve the difficulties encountered.

The early machines were of the type originally known as "briquet hydrogen"—now known as the plunger or dip generators. The next to appear were the dry generators, so-called, in which water was sprinkled or dripped upon the carbide—sometimes now called drip machines. These were an enormous step in advance of the plunging machine, but with the overcoming of the earlier objections other corresponding difficulties were encountered; and later, when the wet process or chute generator was introduced, it was thought that a great step in advance had been made. This opinion was in the main correct, as has been shown by the investiga-

tions of those who have seriously and understand-
ingly studied the problem; but the same thing
happened in this case that occurred when the dry
machine was substituted for the plunger. Where
certain objections were overcome, others were encountered; and while all types of machines, and
especially the latter two, were excellent gas pro-
ducers, they were difficult of management, and some
of them were dangerous. The gradual addition of
parts for protection against each element of danger
or inconvenience has made the machines safer, more
efficient, and durable. At the same time, it has made
them more likely to become disarranged.

The complications which have grown out of the
older and simpler forms of generators have not,
however, rendered possible the charging of the dry
machine with carbide, without the introduction of
air or the opening of the reservoir containing an
explosive mixture of air and acetylene. This, with
such machines, is the greatest fault, and is the one
which has been most difficult to overcome.

With the wet machine, the loss of gas during the
passage of the carbide down the chute, even if the
same be filled with oil, is very considerable when
using finely powdered carbide; and this has not been
overcome. The waste of gas is perhaps not of great
importance; but the possibilities which exist for its
admixture in explosive quantities with air in the room
containing the generator, and the odor which it is
sure to cause, make the generators of either type
most unpleasant companions in the house.

The solution of this problem of getting the carbide into the generator without the admission of air or the emission of gas is of the first importance.

An almost equally difficult problem, but one of less moment, is the removal of the residuum from the machine. The paste or semi-liquid substance which must be removed from the dry machine, and the thin, creamy sludge from the wet machine, are both difficult of management. It has been proposed to wash the residuum into the sewer; but, on the other hand, it has been suggested that the lime would make an insoluble soap with the grease always present in the drains, which would eventually clog them. In country districts, the use of a small cesspool built especially for the purpose of containing the sludge seems a suitable expedient. Indeed, in communities where cesspools are used for the disposal of sewage, there should be no objection to draining directly into them. The lime is an excellent disinfectant, and similar properties are claimed for the acetylene which the water holds in solution.

The author, in the first edition of this book, advocated strongly the use of an out-door generator built within a cistern. Those which he constructed upon this plan for his own use and for experimental work have given no trouble, and have been extremely satisfactory. The same, or even larger, losses of gas while the carbide passes down the chute are present in this machine, which must have a water-seal of considerable length; but being remote from the place where the lights are used, it causes no inconvenience.   The expense of the installation, however, is many times that of the portable machine.   There seems no reason why a small frost-proof building should not be substituted for the underground construction.   It has often been proposed to fill the gas-holder with oil, and to protect the generator by means of non-conducting material.   In that case the heat set free by the reaction between the carbide and the water can be prevented from escaping, and would maintain a sufficiently high temperature to prevent the freezing of the generator in cold weather.   The author has been unable to verify this condition, for lack of sufficiently cold climate.

In this latter form of out-door generator, it would be necessary, in the event of using a dry machine, to employ a saturated solution of brine for the exciting fluid.

The abandonment of some of the earlier types of machines, and the dissatisfaction which even the best of the more modern ones have given from these causes, makes it imperative that some radical changes be made before the generator can be declared a complete success.

Means must be provided whereby carbide, coarse or fine, or in any condition in which it usually comes from the works, can be put into the generator absolutely without the loss of gas or admission of air.

Means must also be provided for removing the residuum under the same conditions, and, in the case of the wet machine, without breaking the seal through which the carbide enters the chute.   Difficult as these problems seem, they are both possible of solution; and upon them the most strenuous efforts of the inventor should be expended. Their solution will at once make the use of the generators within houses as safe as the use of kerosene lamps, and will entirely remove the disagreeable elements of odor and lime-stains. Until this is accomplished, the user of acetylene must tolerate its inconveniences for the sake of its incomparable light.

LIGHTING BY ACETYLENE

INTRODUCTION

In an attempt to set forth the facts concerning the development of the kindred industries of calcic carbide production and the generation of acetylene therefrom, it must be borne in mind that both processes, so far as their industrial application is involved, are of very recent date.

Whatever is written about either must be considered as an exposition of the art so far only as it is known at the moment of writing. Revision or even radical change of ideas may be expected from time to time as continued experiment brings to light new facts about these hardly known substances.

A recent serious explosion of liquefied acetylene prompts the writing of what follows, since it would seem that ignorance or neglect of the admonitions of earlier experimenters was to blame for an accident which cannot but prejudice many against the use of a valuable and safe illuminant.

Since the French have been unusually keen in the pursuit of information concerning calcium carbide and acetylene, and have not only devised many machines for generating the gas, and furnaces for producing the carbide, but have made public the results obtained, their books and pamphlets on the subject have been freely drawn upon for details of foreign practice.

As for the American gas generators, a description of types of those which the author has been able to discover from an examination of the United States Patent Office records, direct interviews with the manufacturers or inquiry among the carbide dealers is included ; but, since hundreds are experimenting in this field, and since two or three new generators are patented in this country each week, it is quite impossible to include the most recent ones.

Suffice it to say that, in the very nature of the problem involved, generators may all be divided into three classes, and that any new machine can differ from the existing forms only in matters of detail.

For the production of acetylene it is necessary to bring calcic carbide into contact with water in some kind of vessel from which the resulting gas may be conveyed for use.

Whether the carbide is thrown into water, or the water is poured upon the carbide, is in general terms a matter of indifference, the result being practically the same.

When, however, the renewal of the carbide becomes necessary in order to keep up a continuous supply of gas, complications immediately enter the problem, which becomes still further involved when the removal of the lime resulting from the reaction is attempted.

The generation of a quantity of acetylene is a very simple matter. The devising of means for delivering a constant supply of gas, supplying calcic carbide to the generator and removing the lime therefrom has taxed the ingenuity of some of our ablest inventors.

The ideal machine has certainly not yet been invented, but existing forms are being constantly improved, and at the present writing, the safe, efficient and cheap lighting of houses by acetylene is an accomplished fact.

HISTORY

ALTHOUGH only recently become of commercial importance, acetylene has been known chemically since 1836, when the chemist Edmond Davy announced the discovery of "a new gaseous bicarburet of hydrogen and of a particular compound of carbon and potassium, or carburet of potassium," in the British Association Reports, 1836, pt. 2, p. 62.

In 1861, the German chemist Wöhler prepared calcium carbide by heating a mixture of lime and carbon in the presence of zinc.

Berthelot, in his classic synthesis of a hydrocarbon, passed hydrogen through a receiver in which an electric arc was maintained between carbon electrodes. The hydrogen combined with the vaporized carbon to form acetylene (C_2H_2), which was passed through an ammoniacal solution of copper, where acetylide of copper ($C_2H_2Cu_2O$) was precipitated.

Berthelot, in 1866, obtained carbide of sodium by gently heating metallic sodium in an atmosphere of acetylene. The acetylene, being absorbed, produced C_2H_2Na, and, upon being raised to a red heat, the hydrogen was driven off, leaving sodium carbide (C_2Na) as a heavy, dark, stone-like mass, which gave off acetylene when thrown into water.

Others since then have formed various carbides,

4

which have been used in the laboratory as á convenient means of producing acetylene.

In a note to the Académie des Sciences, presented
on December 12, 1892, Moissan, who had been experimenting with the electric furnace, made the
statement that "If the temperature of the electric
furnace reaches 3,000° the material of the furnace
itself, the quick-lime, melts and runs like water. At
this temperature the carbon rapidly reduces the
oxide of calcium and the metal is set free in abun-
dance. It combines easily with the carbon of the
electrodes to form carbide of calcium, fluid at this
heat, which is easily recovered."

In March, 1894, Moissan presented to the Académie a sample of pure crystalline calcium carbide
which he had obtained by submitting a mixture of
powdered lime and carbon to the action of the
electric furnace.

Prior to this time, in America experiments had
been conducted with the electric furnace on a very
extensive scale. The Cowles furnace was being
used for the production of aluminum, and carborundum produced in the electric furnace had become
an article of commerce. Mr. Thos. L. Willson had
built a plant containing an electric furnace, with
which he was experimenting upon further improvements in the production of metals and alloys.

According to the most authentic reports, the discovery by him of calcium carbide was an incident
in his search for a flux which should prevent the
spattering of the molten alloys in the furnace from

interfering with and short-circuiting the arc. The use of lime and carbon for this purpose naturally resulted in the production of calcium carbide. While others, no doubt, produced unwittingly the same substance, none had noticed its character or taken advantage of the future which it promised until Mr. Willson was impressed by its possible commercial value.

The French unanimously accord the discovery of the crystalline carbide to the chemist Henri Moissan, and urge most ably the validity of his claims to the discovery of the substance.

The whole matter has been taken up by the *Progressive Age*, of New York, in a series of articles published during the summer of 1898, in which the claims of both Moissan and Willson are very fairly set forth. The bulk of the testimony would seem to show that Moissan had produced the carbide in small quantities in a laboratory furnace prior to its discovery by Willson; but had it not been for American foresight and enterprise, the commercial importance of calcium carbide as a factor in lighting would probably for a long time have remained hidden in the foreign laboratories. There is no doubt that to Willson belongs the credit of the discovery of calcium carbide in its commercial form, and for the industrial purposes of gas-making.

It is not to be wondered, when we consider the possibilities opened up by the electric furnace, that many experimenters hit upon the same facts and were experimenting along the same lines; and it is only charitable to believe that the claims of each inventor are based upon his honest belief.

Since the first edition of this book was written, many other experimenters have been discovered who, between the years 1836 and 1894, made some contribution to Science concerning the properties of acetylene. None, however, conceived the possibility of using the gas industrially until the electric furnace had become a factor of modern manufacturing arts; and while such honor is due them as the chemist usually receives for his discoveries, their work, viewed from an industrial stand-point, is of little importance.

The investigator who wishes to decide for himself the relative merits of the claims for priority of invention of calcium carbide is referred to Perrodil's text-book on acetylene and calcium carbide, a translation of which was published in *Progressive Age*, and the numbers of *Progressive Age* in which the claims of Willson are set forth.

DANGERS OF ACETYLENE

Two or three years ago, when the commercial production of acetylene was first attracting general attention, the most diverse and exaggerated accounts of its dangerous properties were published. By some it was declared to be intensely poisonous, by others frightfully explosive, while a third faction announced that it readily formed explosive compounds with the metals of the pipes and gasometers necessary for its use.

Since the truth concerning this gas is now well known, it seems needless to enter upon a detailed description of the dangers then predicted, and it will be sufficient excuse for the misinformation so generally disseminated to say that much of the early carbide was very impure.

No doubt the gas resulting from its decomposition contained all kinds of undesirable substances, some explosive, some toxic, and some tending to favor the formation of the detonating acetylides of metals.

Now, however, that we may be sure of the purity of our carbide, the dangers attending the use of acetylene have been exactly determined.

Without going into the details of the experiments conducted by the ablest chemists of the time, or re-

8

lating the experiences of those who have labored to bring into vogue the practical and industrial application of acetylene, it will suffice to present a synopsis of the dangers attending the use of this gas and the means for obviating them :

EXPLOSIVENESS.

Acetylene, as alarmists are so fond of stating, is an endothermic substance, which means that in its production a certain quantity of heat is absorbed and disappears. This heat, or its equivalent in some other form of energy, exists in the substance, tending and striving to reassert itself upon provocation. For this reason the gas, under certain conditions, is an unstable body. It tends to resolve itself into its component elements, carbon and hydrogen, and when such dissociation is by any means brought about without the presence of other substances, the result is a certain quantity of hydrogen and a mass of finely divided carbon.

This dissociation can be effected in a body of acetylene gas at atmospheric pressure by the detonation therein of a small quantity of fulminate of mercury or other violent explosive, AND BY NO OTHER MEANS.

Heat, flame, the electric spark or the electric arc itself will not, according to the most careful experiments, produce an explosion under these conditions.

When the gas is condensed, however, it becomes,

with each increment of pressure, more and more unstable, and, in consequence, more easily exploded until the point of liquefaction is reached, when it becomes as dangerous as the high explosives.

The disastrous results which have invariably followed the attempts to liquefy acetylene should be sufficient warning against this procedure.

The only place where there is any excuse for compressing acetylene is in the chemical laboratory, where its properties are being studied by those skilled in dealing with unstable compounds, and where explosions are expected and provided for as a matter of course.

Taking it for granted that, until the dangers attending the use of liquid acetylene have been overcome, the gas under a pressure only slightly above that of the atmosphere will be used in machines for industrial lighting, we may state that, so long as acetylene is unmixed with any other substance, it cannot be exploded by the means usually at command.

An admixture of air with acetylene at once alters the case, for then we have the same conditions which determine the explosion of a mixture of the ordinary illuminating gas with air.

The two gases then behave in nearly the same manner, becoming more and more inflammable as the proportion of air increases until the mixture contains about one part of gas to four of air, when it becomes explosive.

The mixture remains explosive until the proportions are one of gas to twenty of air, after which the dilution is too great for the propagation of flame.

In each stage the acetylene mixture is somewhat more dangerous than the house-gas mixture, simply because its explosion in each case is rather more violent. With a properly constructed generator, however, there should be no chance for the admixture of air with the gas in any proportion, no matter how small and apparently harmless; and, since many generators obviate entirely the admission of air to the system at the same time that they fulfil all the other requirements of successful gas-production, there can be no excuse for taking the slightest risk on this point.

ACETYLIDE OF COPPER.

The presence of ammonia in the gas favors the formation in the gas fixtures of this explosive salt. Well-washed gas should not combine with the small proportion of copper found in the ordinary fixtures. The substance is not easy to make, even in the laboratory, and the amount which could by any chance form in the fixtures is very small.

The use of copper should, of course, be debarred for any part of the system, and especially for the generator or gasometer.

TOXIC PROPERTIES.

A most elaborate series of experiments conducted in France upon men and various lower animals has shown conclusively that acetylene is slightly less poisonous than the ordinary coal-gas in general use.

In experimenting upon dogs, it was found that, when the animals were removed from the influence of acetylene before they had been fatally poisoned, recovery was more rapid than when they were subjected to the effects of ordinary illuminating gas under the same conditions. An examination of blood samples taken every few moments showed that acetylene was rapidly eliminated from the system.

It was also found that fatal results were not produced by the prolonged inhalation of acetylene and air mixtures unless the gas existed in the proportion of more than twenty per cent.

The author may say that, while he is unusually susceptible to the effects generally produced by inhaling noxious gases, he has experienced no inconvenience whatsoever from breathing day after day an atmosphere rich in acetylene.

The danger to be apprehended from leaving the tap of a burner carelessly turned on is too remote to require serious consideration, since the leakage of half a foot of acetylene an hour would require, in an air-tight room eight feet square by eight feet high, fifty hours to produce a mixture of only five per cent. of gas.

The odor of acetylene is so peculiar that a very small leak is quickly noticeable. The odor, which is quite indescribable, is decidedly unpleasant, reminding one somewhat of garlic, or onions.

The products of the perfect combustion of acetylene consist solely of vapor, of water, and carbonic acid. In the case of incomplete combustion, in addition to these products, carbon monoxide, carbon, and hydrogen are produced.

The latter statement is equally true of any of the combustible illuminants, but the amount of carbon monoxide given off from the acetylene flame of the standard burner will be only one-tenth part of that from ordinary gas.

Experiments similar to those mentioned above have shown that animals are affected in about the same degree by inhaling the products from the combustion of equal quantities of acetylene and illuminating gas.

In the case of neither gas need any fear be entertained of the effects of inhaling the products of combustion of burners used for lighting. Cases of injury from this cause have happened, indeed, with illuminating gas, but only when used in large quantities, for heating, in a gas stove without proper provision for ventilation.

EFFECTS UPON THE EYES.

When the incandescent electric lamp came into use there was a general complaint that it "hurt the eyes." Later, the Welsbach mantle suffered under the same imputation.

One rarely hears either blamed for eye injuries at the present day.

The intensely bright light of acetylene will certainly be more than either of the others the object of a similar complaint.

A careful observation of the advent of all these lights has led the author to believe that, so long as a light is a novelty, and so long as individuals, prompted by curiosity, continue to look directly at the flame or other source of light, they will quite naturally be temporarily dazzled and partially blinded.

As soon as the novelty wears off and they are content to look at the objects illuminated, the complaint ceases to be heard.

Man for some thousands of years has had for his type of light the sun, and it is without doubt true that the sunlight is yellow. He takes most kindly to a yellow light, which is the reason the electric arc is so unpleasant, with its bluish tint and moonlight effect, and also the reason that the Wellsbach seems green to most of us. As a matter of fact, the acetylene flame is very like sunlight, and its effect on the eyes cannot but be beneficial, on account of its perfect steadiness.

The only disagreeable feature is, that from its small size, it casts a rather sharp shadow, which makes it unpleasant when burned without a diffusing globe or shade.

ELECTRIC FURNACES

EVER since the beginnings of chemistry, the followers of that science have sought means for producing intense heat.

A charcoal fire urged by bellows or an alcohol flame intensified by the blowpipe, was used for this purpose by the early experimenters.

Then came, with the advent of illuminating gas, the Bunsen burner and the blast lamp in various forms as an important step in advance, and finally the oxy-hydrogen jet with which platinum and iridium could be fused.

Each improvement led to increased knowledge of the more refractory substances, but investigators still longed for a source of heat, which not only would be more intense, but which should be free from the disadvantages of a highly oxygenated flame. The extreme temperature of the electric arc had long been known and had been utilized in a small way in researches upon refractory materials in conjunction with spectroscopic analysis. During the past few years, however, the production of electric currents of immense quantity has become an established industry, and it has been possible to so magnify the small arcs of a decade ago, that an entirely new and most important piece of apparatus has been

developed. The result is seen in the electric furnace, which may vary in size from a small crucible,
in which is maintained the arc of an ordinary street
light, to those huge creations of the carbide works,
where a thousand horse power of energy is converted into the sun-like radiance which fills the space
between the carbons.

The electric furnace consists, in its simplest form,
of a crucible of refractory material, within which an
electric arc may be maintained between the ends of
two carbon electrodes, which enter the crucible for
that purpose.

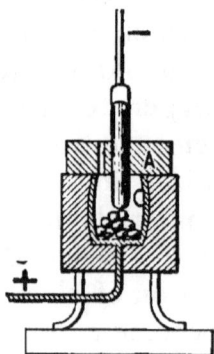

FIG. 1. Siemens Furnace.   FIG. 2. Moissan Furnace.

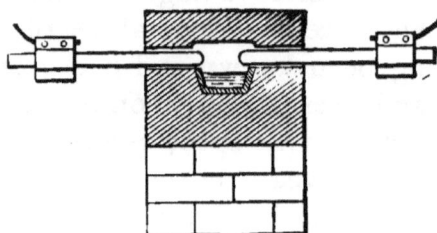

The substance to be treated, generally in the form
of grains, or in powder, is placed in the crucible in
such position that it may be traversed by the electric arc, to whose intense heat it is subjected. The
electrodes are sometimes placed in a horizontal position, sometimes vertically, and, again, inclined.

When horizontal, they may enter the crucible

through holes in its wall, or the crucible may be so shallow that they pass over its upper edge.

When vertical, the lower electrode may enter the bottom of the crucible through a hole, or may consist simply of a block of carbon laid on the bottom, or the bottom of the crucible, or the crucible may itself be the electrode, provided it is a conductor of the current, while the other electrode de-

FIG. 3.   Moissan Furnace.

scends through a hole in the cover when a closed crucible is used, or is guided centrally of the crucible by external mechanism in the case of an open furnace.

The Siemens furnace, Fig. 1, has a graphite crucible embedded in a mass of refractory material intended to prevent radiation. The crucible forms the lower electrode. There is a vent for the gases.

The upper electrode is operated by a magnified arc lamp mechanism.

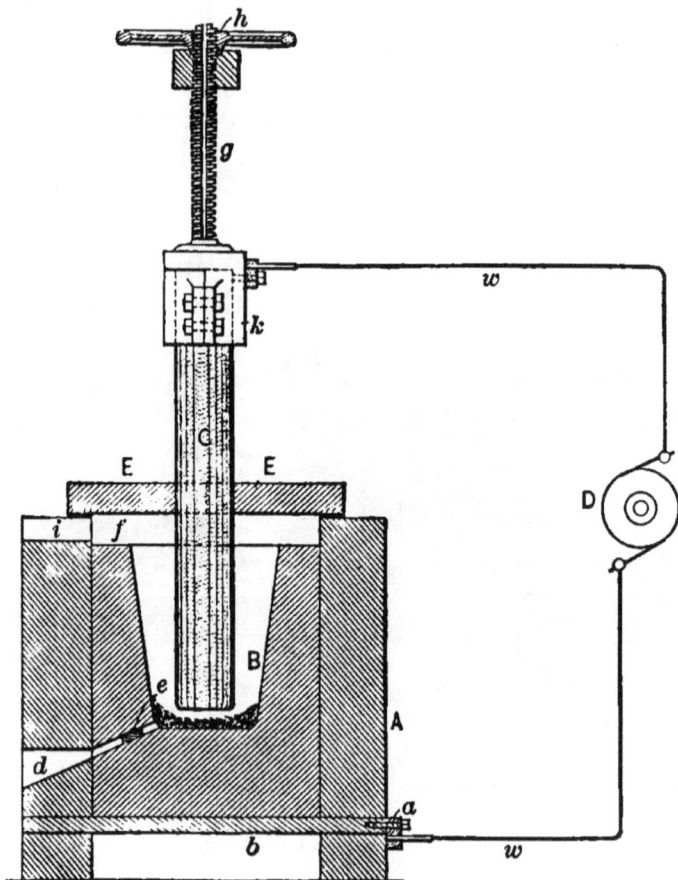

FIG. 4. Willson's Furnace.

The first Moissan furnace, Fig. 2, has horizontal electrodes manœuvred by hand. The crucible is so shallow that the electrodes pass over its upper edge.

The second form of Moissan furnance, Fig. 3, is like the first except that a hole is cut at right angles to the electrodes entirely through the furnace, near the bottom of the crucible. By inclining

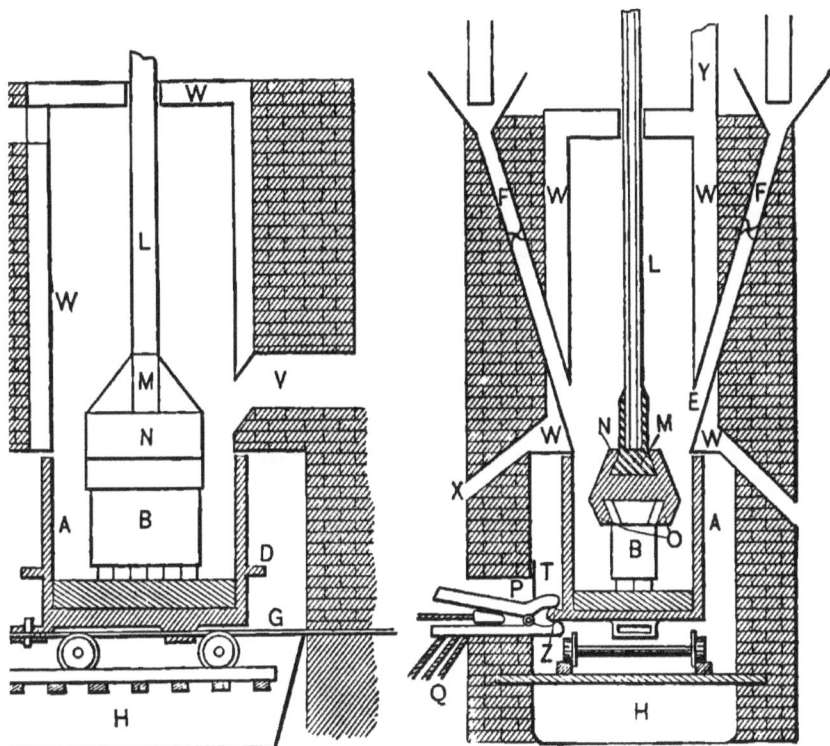

FIG. 5. King Furnace.

the furnace, it may be made continuous in action. The charge is fed in at the upper end of the crosshole, and, after passing through the crucible, the product issues at the other side.

Willson's furnance, Fig. 4, has an outer casing

of brick (A), a crucible of carbon (B), a lower elec-trode of broken carbon, an upper electrode (C), with a wheel and screw h, g, for moving it, a tap-hole (D), and an iron base-plate, to which one pole of the generator is connected.

The King furnace, used at the carbide works at Niagara, N. Y., is on a more elaborate plan, although the additions to its mechanism are of importance only in giving ease of charging and removing the product. The crucible is contained in a small iron car, which may be run out on a track when desired and another substituted for it without loss of time. Suitable chutes allow the lime and coke mixture used for making the calcic carbide to be delivered to the furnace; flues carry off the gases of combustion. The car carrying the crucible is given a backward and forward motion during the action of the current, in order to distribute the contents and to make the action of the arc uniform. The upper electrode is formed of a number of carbons clamped into a massive connector.

King & Wyatt have patented a process for forming calcium carbide, in which, in lieu of an electric furnace, the mixture of lime and coke is placed in a heap on an iron plate which rests on the ground and forms the lower electrode. The upper electrode is supported on a light crane and is lowered down through the centre of the pile.

The carbide forms as a nugget in the centre of the mixture, from which it is removed by means of a pair of tongs.

Some furnaces are provided with movable bottoms for dumping the charge. Others have a tap-

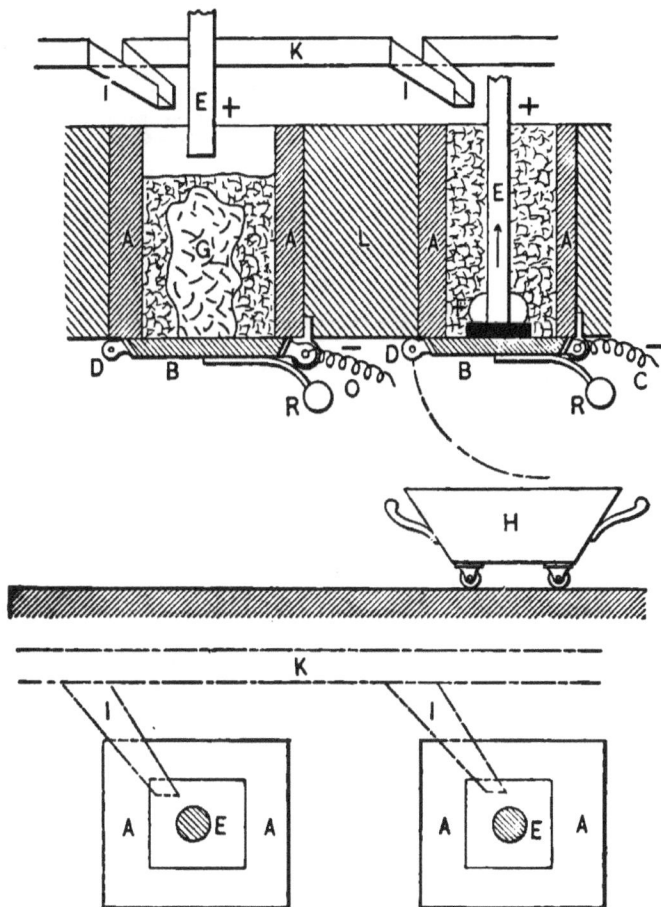

FIG. 6. Bullier Furnace.

hole for drawing off the molten carbide, but the best practice seems to consist in starting the furnace with the arc at the bottom, raising the electrode from time to time, and allowing the carbide to build up in the shape of a block until a considerable thickness is obtained. The current is then shut off, the upper electrode drawn out of the furnace, and

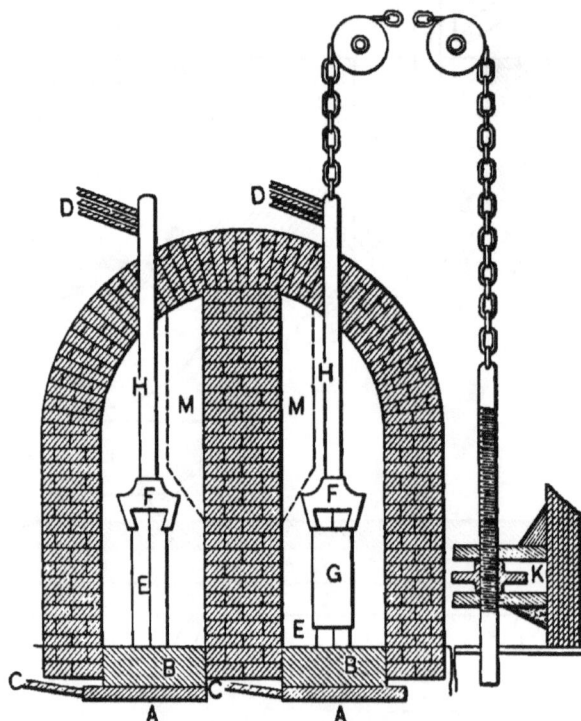

FIG. 7.   Furnace used at Spray.

the crucible removed for cooling, while a fresh one is put in place and the electrode lowered.

The Bullier furnace, Fig. 6, is one having a dumping bottom. The sides are vertical, of fire-clay. The iron bottom serves as the lower electrode. The upper electrode is, as usual, of carbon, which

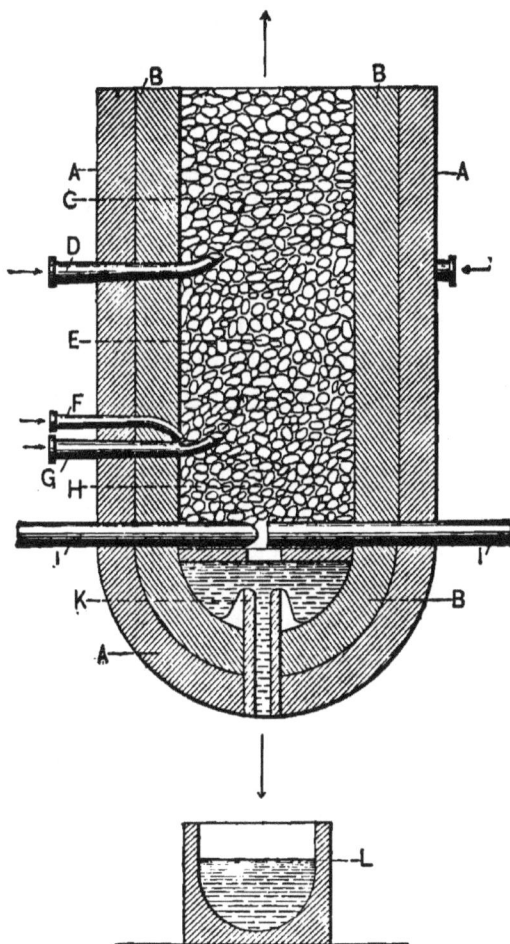

FIG. 8.   Pictet Furnace.

penetrates the centre of the mass of lime and carbon
contained in the furnace.   As it is raised, there is
formed about its end a cavity, into which the contents of the furnace fall, little by little.   The block

of carbide, which occupies the centre of the mass at the end of the reaction, is dropped into a car by opening the bottom of the furnace.

The furnaces used in the carbide works at **SPRAY,** Fig. 7, are of the Willson type, but are double, and are covered by an arched flue, through which the gases escape. The carbons are composed of six blocks, each four inches square and a yard long, held in a clamping head and bound together by an iron sheath. In this arrangement, each furnace must be allowed to cool before the calcium carbide is removed.

M. Raoul Pictet has proposed a furnace, Fig. 8, in which the mixture of lime and coke is acted on, first, by a current of heated air at D, then by an oxy-hydrogen flame, G, as it reaches a lower level, and finally by the electrodes I, I, which melt the carbide. A hole in the bottom of the furnace allows the product to drop through into a receptacle, L.

This is, apparently, an unsatisfactory method, because the coke must be in excess in order to compensate for that burned out of the mixture, and the ash which results from the combustion materially interferes with the proper formation of the carbide.

Several attempts have been made to produce a continuous furnace, but, as yet, with unpromising results.

The continuous process would certainly be a gain in the economical production of carbide, if it could be made to work successfully.

A small portion only of the coke and lime would

be under action at a time, which, as soon as converted, would be automatically removed from the furnace ; the necessity of maintaining the mass of carbide at a high temperature until all was converted would be obviated, and the loss of time in recharging furnaces and waiting for the carbide to cool would be avoided.

A CONTINUOUS ELECTRIC FURNACE

SINCE the foregoing chapter was written, Mr. C. S. Bradley, of New York, has patented a continuously-acting electric furnace, which seems to satisfy perfectly the conditions of uninterrupted carbide production.

The description of this furnace (Figs. 9 and 10), and its operation is taken from the patent specification at considerable length, because, in a general way, it is an excellent account of the manner in which carbide is produced :

"The object of the invention is to permit a continuous and uninterrupted operation of the furnace, and withdrawal of the product, and to protect said product from the action of the air when at a high temperature.

"The furnace is especially designed for employment in the manufacture of metallic carbides. It comprises a receptacle for the charge to be operated upon, in which it inserts an electrode, means being provided for continuously moving the receptacle with relation to the electrode so as to bring fresh portions of material under the action of the electric current. The construction which it is preferred to employ comprises a rotary wheel or annulus, into which projects at one side an electrode, and provided with means for preventing the material from spilling, and means for supplying fresh material to be acted upon by the current, and facilities for removing the product, the whole being so arranged that the operation may be carried on in an uninterrupted manner, the furnace constantly forming fresh additions to the product and permitting the latter to be removed as frequently as may be necessary. The wheel is preferably turned by power-driven machinery, and is provided with a hollow periphery, to which are attached over an arc covering the lower part of the wheel buckets forming throughout said arc a closed receptacle for the material to be operated upon. Said buckets are arranged to be withdrawn or opened when they reach the discharge-end of the wheel-arc. The material, in the form of powder or granules, is supplied to the side of the wheel which contains the electrode or electrodes. The electric arc, or the limits of the space within which the electric action on the material takes place, are wholly within the mass of pulverized material, so that a wall of unchanged or unconverted material will surround the product of the furnace, and the motion of the wheel in such direction as to surround the converted material by a body of unconverted material, and thus exclude air until the converted mass has become sufficiently cool to permit its removal and further treatment for packing for shipment or storage. In the formation

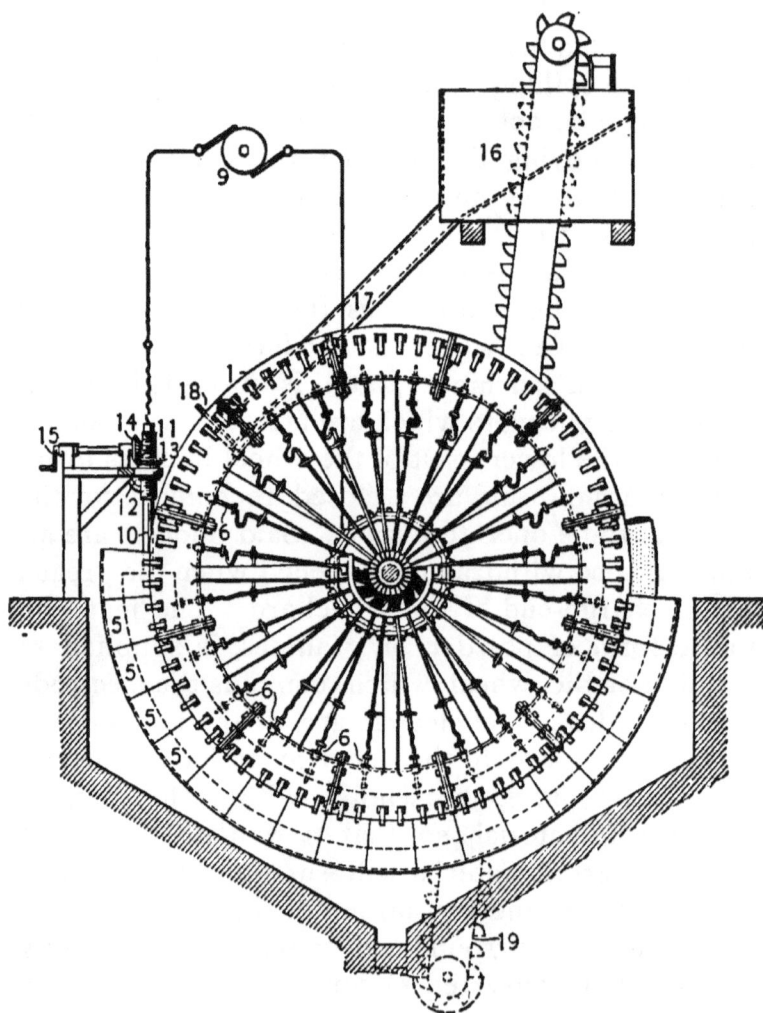

FIG. 9.

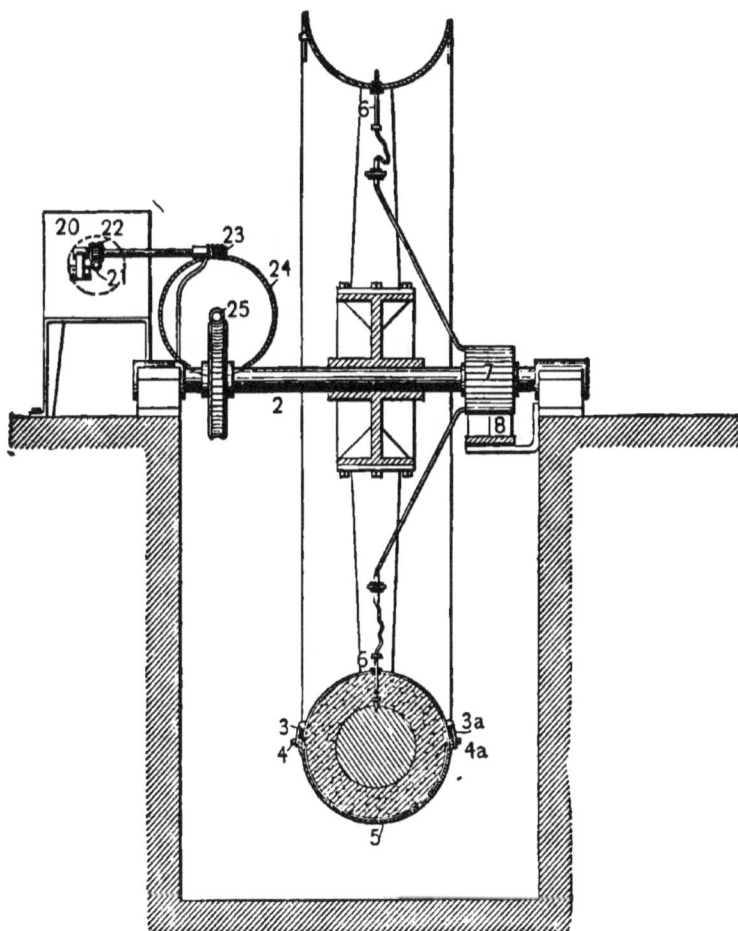

FIG. 10.

of a carbide of calcium, for example, an intimate mixture of ground lime and ground carbon is supplied to that side of the wheel-arc into which the current is introduced, and is fused, permitting the carbon and calcium to combine, and forming a pool of liquid carbide of calcium within the wheel-rim, which pool is surrounded by a mass of uncombined mixed carbon and lime, which acts as an efficient heat-insulator, keeping the walls of the receptacle comparatively cool. As the wheel turns, the pool is withdrawn from the neighborhood of the electric arc, or region of electrical activity, and the liquid carbide cools and solidifies under a superincumbent and surrounding mass of material, which prevents access of air and thus prevents wasteful consumption of carbon by combustion. Thus a core of solid carbide of calcium is formed within a granular or pulverized mass of material, said core growing in length as the receptacle recedes from the electrode until it emerges from the other end of the wheel-arc, when the removable sections of the wheel-rim may be taken off one at a time, permitting the pulverized material to fall away from the solid core of carbide, which may be broken off or otherwise removed periodically. Thus the formation of carbide goes on continuously without necessary interruption for recharging or removal of the product.

"Fig. 9 is a sectional view on a plane at right angles to the wheel-axis. Fig. 10 is a sectional view on a plane parallel to the wheel-axis.

" 1 represents a wheel formed in sections and

bolted together, and having a horizontal axis mounted in boxes at or near the floor-level. The rim of the wheel is concave in cross-section, and is provided at intervals with pivoted latches (3, 3ᵃ) to engage studs (4, 4ᵃ) on semi-cylindrical sections of plate-iron (5) to support them on the wheel. Auxiliary plates of thin sheet-iron may be bent around the joint between the sections on the inside of the wheel-rim, to prevent the pulverized material from sifting through the cracks at the joints. The wheel may with advantage be made about fifteen feet in diameter, and the rim and plate-iron sections of such proportions as to form a circular receptacle of thirty-six inches in diameter. The inner wall of the wheel-rim is provided with holes at intervals to receive copper plugs (6) connecting with the several plates of a commutator (7) by conductors (6ᵃ), on which bears a brush (8) connecting with one pole of an electric generator (9). The other pole of the generator connects with a carbon electrode (10) about four inches in diameter mounted in a sleeve (11) provided with a screw-thread on the outside, which engages an internally threaded sleeve (12) secured to a bevel-gear (13) meshing with a gear (14), on the axis of which is a crank (15) for adjusting the electrode. The electrode and its regulating mechanism are mounted on a framework adjacent to the wheel-pit, so that the electrode may be fed into the receptacle formed by the wheel-rim and the rim sections when partly consumed.

" 16 is a feed-hopper provided with a spout (17)

projecting into the wheel-rim, and a gate (18) for regulating the supply of mixed material to be acted upon.

"The wheel-pit is preferably provided with sloping sides, so that any powdered material which drops from the wheel, at its discharging end or elsewhere, may slide by gravity to a conveyer (19), the buckets of which return it to the feed-hopper, to again pass through the furnace.

"The wheel is preferably connected with an electric motor by speed-reducing gearing. Said motor is shown diagrammatically at 20. The motor-shaft carries a worm (21) acting on a spur-gear (22), on the shaft of which is secured a worm (23) meshing with another gear (24), on the shaft of which is a third worm (25) meshing with a gear on the wheel-shaft. By this mechanism, a very slow speed of the wheel may be maintained, a complete revolution being made once in five days. In using the apparatus, the rim-sections are latched over the wheel-rim over an arc covering the lower part of the wheel, and the gate of the feed-hopper is opened. A charge of intimately mixed pulverized carbon and lime, in proper proportions to form carbide of calcium, falls into the receptacle around the wheel-rim and accumulates until the top of the electrode is immersed therein. The circuit of the dynamo-electric machine may then be closed and the electric motor thrown into operation. As the charge is moved away from the electrode, intense heat is created and the refractory material fuses, forming a pool of

liquid carbide of calcium, or other compound, de-
pending on the nature of the furnace-charge. As
the wheel turns, the pool gradually recedes from the
electrode and slowly cools while inclosed within
walls of refractory, uncombined material on all sides,
the cool product forming a bottom for the liquid
compound.   Thus a continuous core of the product
is formed, new rim-sections being added by a work-
man at intervals of a few hours.   The electrode, at
starting, should project well into the receptacle, and,
as the wheel turns, the electrode rises relatively to
the charge, and, when it reaches a point near the
top of the rim-section, a new rim-section is hung on
the wheel by means of the next set of supports, and
a strip of sheet-iron is bent around the joint between
the rim-sections.   The gate of the hopper is then
opened and the rim filled, or partially filled, with
material.   As this material in its powdered state is
a very poor conductor of electricity as well as of
heat, the immersion of the electrode does not inter-
fere with the heating action.   When a new rim-section is added on the electrode side of the wheel, one
is removed at the other side.   Thus the process
continues until the solid core of the furnace product
appears at the discharge-end of the wheel, when a
rim-section is taken off and the powdered material
falls into the pit, leaving a pillar of solid product
projecting vertically, which may be broken off or
otherwise removed.   Solid carbide of calcium is a
conductor of electricity, and the copper plugs make
a good contact with the same, thereby constituting

the carbide itself one of the electrodes. The action of the commutator leads the current to a point of the carbide core close to the electrode, and thereby prevents unnecessary resistance, which would intervene if the plugs were more widely spaced. The conducting plugs (6), which are remote from the arc, help to carry the current, and thus heating of any one contact with the carbide core is reduced.'

GENERATION OF ACETYLENE

THE calcium carbide of commerce comes to us in air-tight cans of various sizes. The usual package holds either fifty or one hundred pounds. Upon opening the can, we find a heavy, dark-colored, stone-like substance in lumps of various sizes. The largest pieces are the size of one's fist, while the smallest are in the form of grains broken from the larger pieces in shipment and travel. The specific gravity of calcic carbide is 2.22. Its fracture presents a crystalline surface like that of broken cast-iron, but almost immediately loses its lustre when exposed to the air, becoming covered with a film of lime.

Upon dropping into a tumbler of water a piece of the carbide as large as a hickory-nut, a surprisingly violent reaction takes place. Acetylene is rapidly generated at the surface of the carbide, and, rising in the form of bubbles, throws the water into violent ebullition. A considerable amount of heat is liberated at the surface of reaction, which may boil the water in the tumbler if the piece of carbide is large.

The whole mass of liquid, which is rapidly whitened by the lime set free, is in such violent commotion and is so *soufflé* by the issuing gas that many bubbles carry away sufficient heat to break as little puffs of steam before the water has reached the boiling-point. The issuance of steam must not, however, be confounded with the appearance of the first bubbles which arise, and which being charged with finely divided lime give off a white dust which looks much like condensed steam. If this experiment is repeated upon a larger scale, with several pounds of carbide and a pailful of water, the phenomena are more striking. In this case the nature of the vapor. which rises soon after the reaction begins, may be discovered by noting that it does not disappear like steam, and it will be found, upon putting the hand into the bubbling water, that the liquid which seems to be boiling is quite cold.

The violence of the reaction is surprising, even after seeing it many times. The water during the reduction of the carbide is in a state of the most violent commotion. Bubbles rise in great masses to the surface of the liquid, but there instead of bursting, as one might expect, they roll over and over among themselves, and seem to the eye to go down again into the places whence they came. As the carbide is exhausted there is a gradually lessening disturbance of the water until the end of the reaction, when a thin stream of individual bubbles rises to the surface where each may be seen to break in the way one might expect.

The water, after the reaction has ceased, is left white and thick with lime, which gradually settles to the bottom of the containing vessel. If the water is then decanted the lime remains in the form of a semi-fluid paste, which flows easily. A few pieces of impure carbide, together with an occasional nodule of glistening carbides and silicides of various metals may be found at the bottom of the fluid. The latter are formed in the electric furnace from the various impurities contained in the lime and coke.

When calcium carbide and water are brought together, a very simple mutual disintegration takes place. The reaction results in two new substances, namely hydrated lime and acetylene.

Chemically expressed the formula is:

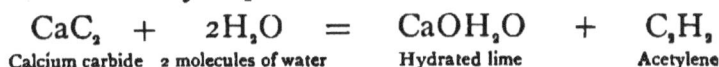

$$CaC_2 + 2H_2O = CaOH_2O + C_2H_2$$
Calcium carbide 2 molecules of water   Hydrated lime   Acetylene

Considering the molecular weights:

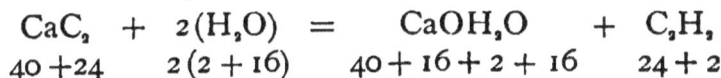

$$CaC_2 + 2(H_2O) = CaOH_2O + C_2H_2$$
$$40 + 24 \quad 2(2 + 16) \quad 40 + 16 + 2 + 16 \quad 24 + 2$$

and reducing these to decimal relation we find

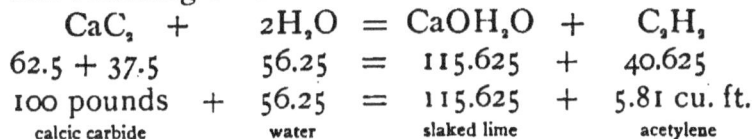

$$CaC_2 + 2H_2O = CaOH_2O + C_2H_2$$
$$62.5 + 37.5 \quad 56.25 = 115.625 + 40.625$$
$$100 \text{ pounds} + 56.25 = 115.625 + 5.81 \text{ cu. ft.}$$
calcic carbide   water   slaked lime   acetylene

A certain quantity of heat is set free during the generation of acetylene, which in the method described is manifested by a rise in the temperature of the water into which the carbide is dropped.

A more striking way in which to show the liberation of heat that attends the decomposition of calcium carbide, is to dip a small piece, held between the thumb and finger, into water and then to quickly withdraw it therefrom. As the moisture reduces the carbide, the temperature rises to such a degree that the piece must soon be dropped. When water is supplied to a considerable portion of calcic carbide, drop by drop, or in such small quantity that the carbide is not wholly submerged, the temperature of the inner portions of the mass may rise several hundred degrees, even reaching a red heat. If a wooden pail is filled with carbide and a small quantity of water poured on the mass, the pail may take fire after an hour or more.

Under these conditions the acetylene is given off with clouds of steam and vapor, the gas has a strong odor of ammonia, and the reaction seems more violent than in the case where the water was in excess.

The total amount of heat liberated is of course the same in each case, provided equal weights of carbide are used ; but in the first instance the water which surrounds the carbide is put into violent motion by the rising gas, so that every ounce of it is successively brought into contact alternately with the carbide and with the walls of the containing vessel, where its heat may be conducted away. Besides this, the specific heat of water is so great, that its temperature is raised less than that of other bodies by the same cause.

So long as the carbide is surrounded by freely moving water, it is doubtful if its temperature could rise to a point at which the gas would be decomposed, for water at the boiling-point has remarkable heat-absorbing capacity, provided its circulation is maintained.

When the carbide is in excess, however, the water which is supplied forms a paste with the reduced lime, and surrounds the carbide with a non-conducting coating. The issuing gas affords the only exit for the heat, and not only comes off at a high temperature, but is often partly decomposed while the unreduced carbide increases constantly in temperature.

Moissan has shown that when the liberation of acetylene is attended by a considerable rise of temperature the gas is partly decomposed into substances having a similar structure, known as polymers, such as benzine, styrolene, anthracene, tar and naphthalene.

Acetylene is then said to polymerize, and the gas is partly changed into benzine vapor. If the temperature rises still further the elements of the gas are separated; hydrogen is given off, and carbon, in a state of fine powder, is set free.

The simplicity of the process of making acetylene from carbide would lead one to suppose that the construction of a generator for supplying the gas in the quantities required for lighting buildings would be an easy matter. Such, however, is not the case, and indeed so great are the difficulties encountered in producing a generator which may be depended upon for an unfailing and uniform supply of pure gas, that although hundreds, nay thousands, have labored upon the problem, it is safe to say that to-day not a single generator can be purchased which may be called a perfect apparatus.

The conditions to be satisfied in an ideal generator are as follows:

First—Safety:

A—From explosion of confined gas due to presence within a closed generator.

B—From fire risks.

Second—Economy:

A—In the yield of gas, whereby the maximum output of acetylene is obtained from each pound of carbide, and where all waste in using or recharging the generator is avoided.

B—In the production of a pure gas free from polymers or other products of decomposition, which lower the candle-power per cubic foot.

Third—Ease of management: simplicity.

Fourth—Non-liability of the apparatus to get out of order when put in the care of unskilled persons.

The construction of a generator possessing in full all these requirements is a difficult matter. So far as is known it has not been done. It is proposed to discuss the problem point by point, but it will first be necessary to describe the lines along which experiments have progressed and upon which existing generators have been built.

Acetylene has hitherto been made by either the " dry process " or the " wet process."

THE DRY PROCESS GENERATOR

In the "dry process" generators a large quantity of calcium carbide is placed within a gas-tight reservoir where it is automatically sprinkled with small successive charges of water, the intention being that the addition of water shall be proportional to the yield of gas.

Generators built upon this plan consist usually of a separate generator and gas-holder, so arranged that at each downward movement of the gas-holding bell a small quantity of water will be admitted to the generators in which it will fall in small streams upon the carbide contained therein.

A typical generator of this class is shown on the page following.

The generator proper, A, is an iron box closed by a gas-tight door. Within it is a pan for holding the carbide, B, and a sprinkler, C, fed with water through the pipe, D, and cock, I, from the tank, J.

The generator is surrounded by a bath of cold water, E, which contains a coil of pipe for cooling the gas on its way to the gas-holder, L.

A safety-pipe, T, leads to a water-seal, G, which relieves any accidental rise of pressure of gas in the generator, and also serves as an escape for the water which is condensed upon the cool walls of the generator.

The gas, after passing through the cooling-coil, is

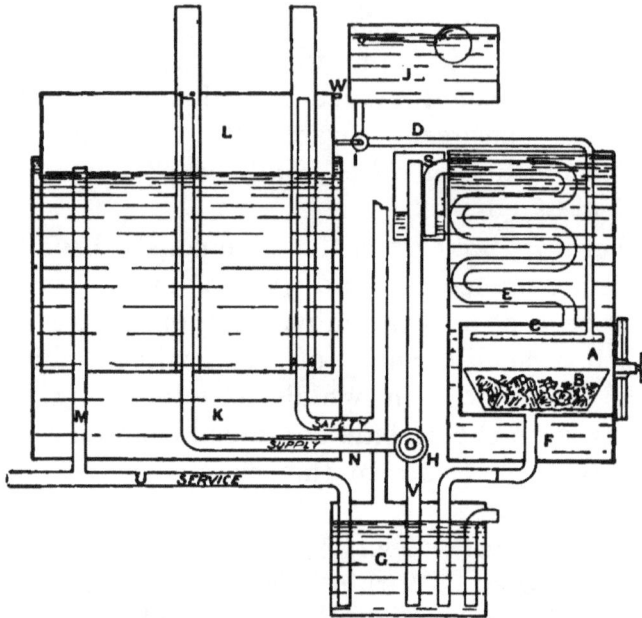

washed by bubbling up through the water in the washer, S, and thence flows down to the three-way-cock, H, where any moisture it contains is drained off to the safety-seal, G, by the pipe, V. This pipe acts in emergencies as a safety vent as well as a draining pipe, and should be of ample size. From the cock, H, the gas passes to the bell, L, of the gas-holder. Thence, it is taken to the burners by way of delivery-pipe, M, past the drainage-tube, U, through the dryer and filter.

The cock, I, is normally held shut by a spring, but is opened by the pressure of a finger, W, projecting from the bell when the latter sinks to a predetermined point.

In some dry generators, such as that of Dickerson (Fig. 21), there is an arrangement whereby the quantity of water entering the generator is definitely measured at each descent of the gas-holder bell. In the "Niagara" generator, a "tip-tank" is used. This fills with water from a constant source, and at each descent of the bell is emptied into the generator.

The gas-holder should be provided with separate inlet and outlet pipes. Moisture is carried into the house-piping when one pipe is made to answer both functions. Either pipe may be made, however, to act as a guide to the bell, or the safety-pipe may be used for this purpose. Some prefer to have an independent guide consisting of a central rod or of a pair of rods—one at each side of the bell, as in Figs. 38 and 26.

The gas-holder should be provided with a safety pipe, which telescopes into a larger tube carried by the bell. When the bell is so full of gas that the orifices near the lower end of the outer tube have risen to the surface of the water contained in the gas-holder, any further supply enters these holes and is carried to the seal G and thence to the outer air, where it is discharged. The gas on its way to the burners should pass through a dryer, in order that moisture may not be carried into the filter nor into the house-pipes, where it may condense and give trouble by cutting off the supply of acetylene, or may freeze and burst the pipes.

A filter is shown at Q, consisting of a cylinder of

loosely woven but thick fabric, through which the gas passes.

This is a really necessary attachment, although rarely used. Without it the finely divided lime, which always comes into the gas-holder with the acetylene, gives trouble by entering the house-mains and eventually clogging the burners. Indeed this is the principal cause of burner stoppages, and for this reason it is also advisable to insert into the base of each jet a small pledget of cotton as an additional filter.

In a plant installed by the author, ten per cent. of the hundred burners used needed cleaning daily, until these filters were used, since which time none has shown signs of clogging.

In order to make clear the other difficulties which may arise with this class of generator, even the ideal one figured herewith, they will be described in detail.

Beginning at the reservoir, J, the water-supply may fail. This may be avoided by connection through a ball-valve with the town supply, whereby the level is kept constant.

The cock, I, which is held shut by a spring, may fail to close or may leak, causing a waste of gas. The remedy for this is to use a "tip-tank" or a measuring cock as in Fig. 11.

The cooling coil may clog with lime or the products of polymerization. In that case gas wastes through the safety-seal, G. The cure is a large coil easy of access for cleaning. Means for flushing the coil with water might be devised.

The three-way cock may leak or become clogged. If, however, the seal, S, is so arranged that it cannot become dry, the cock may be discarded. The seal, however, introduces another difficulty.

When the gas bubbles violently through the seal 'a rhythmic disturbance is set up in the gas-holder which makes the lights near the generator flicker, and sometimes is violent enough to extinguish them. If the seal is abandoned the three-way cock, or its equivalent, must be retained, and *vice versa*.

The retort cannot be entirely filled with carbide. Consequently at each charging a quantity of gas escapes when its door is opened and an equal quantity of air is shut up in the generator each time the door is closed.

The issuance of gas is an element of danger, although not a great one when the generator is intelligently managed. The entrance of air, however, lowers the candle-power of the gas, and may be sufficient in quantity to produce an explosive mixture in the bell.

It certainly does produce an explosive compound in the generator, but the fear that the heating of the carbide, which attends the production of gas in this type of machine, may ignite the mixture, is probably groundless.

The yield of gas from freshly introduced carbide is so rapid, that the air is swept out of the retort before the temperature of the carbide has reached a dangerous point.

The air space is none the less an objectionable

feature in any generator, and should be cut down to the least proportions possible. No way has as yet been discovered for avoiding it in the "dry machines."

The trap, G, may clog with lime to such a degree as to become inoperative as a means for relieving excessive pressure. It may, however, be flushed out occasionally by pouring water into the retort. There is no danger of its getting dry, for all the moisture in the system should drain to it.

The door of the retort may leak, in which event a dangerous quantity of acetylene may escape. A vertical retort closed by a cover having a water-seal prevents this trouble, provided the seal is maintained at a constant level by a float-valve or by constant vigilance.

Gas continues to be generated after the water-supply ceases, in an amount too large for the bell to contain, and is wasted.

When a machine of this character is started with a fresh charge of carbide, little difficulty is met in regulating the water-supply automatically to flow to the carbide, and generate gas in proportion to the amount used.

After being in action a short time, however, the lime disengaged forms a protective and absorbent coating over the blocks of carbide, which takes up the water as fast as delivered until it becomes saturated. It is then only that the water gets free access to the carbide.

Consequently the machine responds slowly to the

addition of separate increments of water. On the other hand, when the supply of water is shut off, the carbide, being hygroscopic, continues to absorb moisture from the wet lime, causing the evolution of gas to continue for a longer time than was intended.

This action causes irregularity in the working of the machine, and when coupled with the fact that the hot lime absorbs a larger quantity of water than it can hold when it has cooled, makes machines operated on this principle unsatisfactory.

The polymerization of acetylene by the rise of temperature in a machine of this class is probably its worst feature. When a large quantity of carbide is charged into the retort, and is then acted on by successive small additions of water, the temperature may rise to the point at which the acetylene is changed as generated into other substances—such as benzine—or may even reach a red heat, at which the gas is split up into carbon and hydrogen. In the latter case both phenomena are present as a rule. Lampblack collects in the water-seal, and benzine may be detected in the gas-holder.

The loss of illuminating power is not the only result of these changes. Certain oily or tar like products are condensed in the pipes, and are carried by the benzine vapors to the burners, where they soak into the lava tips, there to carbonize and eventually stop up the jets. Much of the trouble given by the clogging of the burners is due to this cause.

An index of the amount of decomposition taking

place in the generator is the discoloration of the lime residuum. If it is nearly white or of a bluish tint all is probably going well, but the appearance of red or yellow spots and patches of discoloration indicates that the temperature has been too high.

It is a matter of some doubt, however, at just what temperature acetylene begins to decompose.

The makers of dry machines invariably affirm that the importance of the matter is much exaggerated, while the manufacturers of wet generators aver that it is the most important fault corrected by their plan of operation.

The water-jacket has been abandoned in many dry generators, as being nearly useless in keeping the carbide cool. The carbide is generally placed in a pan which is then slipped into the generator. The water-jacket is, in this case, too remote from the carbide to absorb much of the heat which is set free locally at the point of reaction only.

When a water-jacket is not used, the cooling coil through which the gas passes is sometimes placed directly within the gas-holder.

The dryer through which the acetylene passes on its way to the filter is preferably filled with calcium carbide, which, in addition to absorbing the moisture from the gas, gives up more gas in so doing.

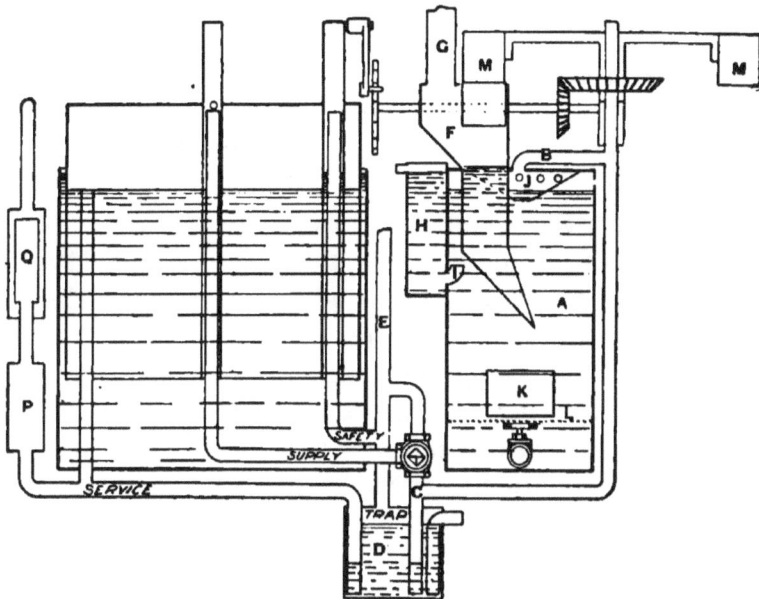

THE WET PROCESS GENERATOR

This is shown diagramatically above, in connection with a gas-holder. The generator consists of a cylindrical can, closed by a top which is pierced with two holes—one for the exit of gas into delivery pipe, B, and the other for the entrance of a chute which depends through the top of the generator to a point considerably below the surface of the water, and is of such shape that a piece of carbide dropped into its upper end will be delivered into the generator at a point where the rising bubbles of gas cannot re-enter the chute nor escape, except by the gas-pipe, B.

While in the dry machine a measured, small quantity of water is periodically delivered to a large mass of carbide, the action in the wet generator is exactly the reverse. The carbide is dropped in small, measured quantities into an excess of water. The generator may be of the very simple type shown in Figs. 36 and 37, in which case the carbide is dropped down the chute whenever the bell has nearly reached the lower limit of its travel. The bell, however, may be made to automatically perform this operation in a number of different ways. In the diagramatic figure, the descent of the bell turns a ratchet-wheel which moves the circle of small carbide cans (M) one by one, over the top of the chute (F), at which point a latch is tripped and the contents of each can is delivered successively down the chute into the generator. The gas which bubbles up rapidly through the water contained in generator, A, passes by way of the pipe, B, to the Tee at C, where any water or moisture which is carried over with the gas drains into the trap, D; the gas then rises through the three-way cock and enters the gas-holder. The arrangement of the safety and delivery pipes is the same as described in the dry machine, both being drained into trap D.

The safety-pipe (E) is carried out of doors, preferably to top of building.

The top of the chute is extended laterally to accommodate vent pipe, G, which should also be carried to the highest convenient point outside of the building. A sharp upward draft will normally exist in this pipe, and will carry away any gas which rises through the chute during the fall of the carbide into the generator.

A filling box is built upon the side of the generator, with which it communicates by a short upturned pipe (I). Water is added from time to time through this box, which also allows the water-level to be seen. A water-glass in this situation soon becomes so coated with lime, that it is useless.

A baffle plate (J) surrounds the entrance of the delivery pipe (B), in order to prevent the entrance of water thereto. This plate is funnel-shaped, having a hole in its centre below the water-line. Water or lime, which enters the side holes in the baffle plate, drains away by the central hole.

At the bottom of the generator is a valve through which the reduced lime in the form of thick cream may be drawn out, and at L is a grating which prevents the unreduced pieces of carbide and other impurities from falling through into the lime-chamber. These are occasionally removed through the hand hole, K.

There are certain drawbacks to the generation of acetylene by this type of machine, and certain accidents which may happen.

The gas in the holder is kept from returning and passing out through the chute (F) entirely by the fact that the chute forms, with the water in the generator, a water-seal.

If, through carelessness in allowing the water in the generator to get low, or by reason of a leak in the generator, the seal at the bottom of the chute is broken, the gas in the holder immediately escapes into the chute, and is thence carried out and wasted through the vent pipe (G). This trouble may be obviated by connecting the filling tank (H) with a constant water-supply, and maintaining its level by means of a float valve, but in that case it seems simply to be adding another element of complication, which in its turn may get out of order.

A leakage in the draw-off cock or hand-hole results in the soiling of floors with the milk of lime which is formed in the generator.

When carbide is dropped into this type of machine the evolution of gas is very rapid, and the surface of the water in the generator is so disturbed by the issuing bubbles that if the water-level is too high, a very considerable quantity of water and lime is carried bodily into the pipe, B, and forced by the rush of gas in some cases as far as the bell of the gas-holder. It is for this reason that a separate delivery pipe is quite necessary, as the moisture or free water will settle in the gas-holder before reaching the delivery pipe. Ordinarily the trap (D) takes care of the water which is thrown into pipe B, providing the latter is of ample area.

Of course, the trap may clog, but by completely filling the generator with water, the pipe and trap may be flushed and the lime driven out.

The three-way cock may leak, but the safety-pipe, which is connected with it, prevents the gas from entering the house. Fortunately, in this class of generators there is almost no rise in the temperature of the carbide, and the gas is very thoroughly washed by bubbling up through the water in the generator.

In case any part of the mechanism which delivers carbide to the generator fails to work, the machine is rendered inoperative and the supply of carbide ceases. This is a good feature in a machine which is somewhat more complicated than the dry machine.

Of course, no air can by any possibility enter the gas-holder, unless the chute-seal is broken, but there is with each charge of carbide a certain amount of gas lost up the chute during the passage of the carbide through the water into the generator. Usually this loss of gas is insignificant, but if the carbide is in very small pieces they sink slowly through the chute, and the loss may be as large as half the amount of gas which would be generated from that charge, and in that case the reaction which takes place in the chute is so violent that the water is carried up in a great mass of bubbles and foam, which sometimes overflows the top of the chute, spattering lime and water freely about in the vicinity of the machine. The only way which has been discovered thus far to prevent this difficulty is to use carbide broken into lumps of fairly uniform size. When it becomes necessary to use finely broken carbide, the boiling over and loss of gas may be prevented by enclosing this fine carbide in a common paper-bag, which may be dropped either by hand or automatically down the chute. In the course of about thirty seconds the paper is sufficiently wet for the evolution of gas to begin; the bag soon bursts open and the gas is given off rapidly.

Unfortunately, the manufacturers will not as yet sort the carbide into pieces of uniform size. This is perhaps the only product of a similar character which cannot be so procured. Almost every other material which is broken for ease of shipment or use, such as coal or stone, may be purchased in pieces of nearly any size desired. It is understood that in the new works at Niagara Falls and elsewhere, the carbide will be put through screens by which it will be separated into different grades.

There is another class of generators, which, although nearly obsolete in this country, was much used in the first days of lighting by acetylene.

In these generators the carbide in considerable quantity is suspended over a large body of water within a gas-tight vessel, and is so arranged that it is alternately immersed in and withdrawn from the liquid.

In some instances the carbide is suspended from the top of a movable bell, as in Fig. 29, whereby it is lowered and raised in turn from the water. In others, as in Fig. 30, the water rises and falls in relation to the carbide. This method of generating acetylene is, perhaps, the worst that could be devised, as it has all the faults of the others with no redeeming traits. It is almost impossible to design such a generator which can be recharged without great waste of gas.

Then the evolution of gas is extremely uneven, being violent at first and rapidly becoming less as the carbide is exhausted.

The carbide is entirely withdrawn from the water after each immersion, so that it is not only removed from its cooling influence, but the moisture taken up with the carbide drains away, leaving it in the best possible condition for generating heat.

The tendency for the water to be thrown into the mass of carbide by the violence of the reaction following its first immersion, necessitates the use of impervious plates which divide the charge into thin layers. Even then the water is thrown up around the sides of the charge, and more gas is given off than can be used or the holder contain. For experimental use, where a small quantity of acetylene is occasionally required, this generator is convenient. It may be very small, is the simplest of any, and may be considerably improved in the uniformity with which it generates gas by covering the water with a thick layer of oil. Under these conditions and where the generator is small, it is very satisfactory for use in the laboratory.

Professor Vivian B. Lewes, the leading English authority on acetylene, sums up the generator situation in the following words:

"The generators of the third class are undoubtedly the best, as with the water kept in excess and charged at the rate of eight pints per hour for each pound of carbide decomposed, it is impossible for the temperature to rise above the boiling-point of water. To ensure this, however, there must be a false bottom to the generator, capable of being occasionally rocked, about a foot to eighteen inches above the true bottom, the lime sludge being drawn off from the bottom, and water admitted *pro rata* from above. If there is no false bottom and a big charge of carbide is introduced, it becomes so coated with lime that the melting-point of tin may be reached in the interior of the mass, but under all conditions this class of generator yields the purest gas, as the acetylene, having to bubble through the lime-water formed in the generator, is washed free from most of its impurities.

" The generators of the first class, in which water is allowed to drip or flow on to the carbide, have several drawbacks. A popular fallacy with the makers of automatic generators of this class is that when the water-supply is cut off the generation of gas ceases, whereas in fact it continues for an hour to an hour and three-quarters longer, although the rate at which the gas is evolved gradually gets slower and slower until it ultimately ceases, this con- tinuation of generation being partly due to conden- sation of water vapor on to the undecomposed car- bide as the generator cools and the decomposition of water mechanically held by the slaked lime, while, if the temperature be very high the slaked lime is itself dehydrated. In this class of generator the temperature frequently rises to a sufficient de- gree to cause the formation of benzine, while in some cases tar is produced.

" Overheating, when it reaches this point, is readily detected by the appearance of the lime left in the generator; this should be practically white, but if it is found to be yellow, serious overheating has un- doubtedly taken place, while the appearance of oil or tar in any part of the apparatus is a still more damning proof of its unfitness.

" The worst offenders as regards overheating are

to be found in the second group, where the alternate dipping of the carbide into water and its removal as the bell rises allows the action to proceed without any cooling influence being brought to bear upon it, and it frequently happens that under these conditions the whole charge of carbide becomes red hot, tar vapors, hydrogen, various hydro-carbons and benzine vapors are mixed with the acetylene, and its illuminating value drops by leaps and bounds until its light-giving properties are no better than those of a good oil gas, while the burner tips become choked and smoking and general discontent reigns supreme."

Returning now to our list of conditions which the ideal generator should fulfil, we shall be able to find what elements are present or lacking in existing machines, and along what lines further development is likely to proceed.

In the matter of safety from explosion due to pressure of gas in the generator, it need only be said that at the present day no generators are made in which a dangerous rise of pressure could take place. Danger from explosion of the gas in the generator by coming in contact with fire is, however, a more real hazard.

In the dry machines the retort must be opened in order to insert fresh carbide. During this operation the gas within the retort escapes, while air gradually takes its place, making both within the retort and in its immediate neighborhood an explosive mixture. In the wet machines the same things happen when the water is drawn off for recharging. Theoretically, this is unnecessary in the latter class of apparatus; but in practice it nearly always happens, and considerable skill is required to prevent its occurrence. In both machines gas may leak from the cocks, and in the wet generator some is always lost during the passage of the carbide down the chute. This may be removed by the ventilator; but, on the other hand, under certain conditions of wind and weather a "back draft" will exist which drives the gas into the room containing the machine.

The clogging by frost of the safety-pipe leading from the gas-holder has been suggested as one source of danger. If the safety-pipe is properly drained, however, this cannot happen, and is mentioned as a caution only against generators which are poorly made.

The installation of any generator should include the construction of a specially ventilated room or closet of sufficient size to contain the complete apparatus, and a small quantity of carbide for immediate use. With this precaution and the living up to the hard and fast rule that the room shall not be entered by artificial light, any modern acetylene generator, be it wet or dry, is, in the opinion of the author, entirely safe as a primary cause of fire.

In the matter of economy and maximum yield of gas, both classes of generators are deficient, with a slight advantage in favor of the wet generator.

Both waste some gas. The dry machine loses

an appreciable quantity at each opening of the re-tort, with the further disadvantage of entrapping a like amount of air. In addition to these channels of waste, there is in the dry generator the ever-present trouble of over-production and waste through the safety pipe ; and in some a considerable degree of polymerization takes place, with its attendant inconveniences of clogged burners and lowered candle-power.

In the wet generator there are two causes of loss in economy. The first results from the time which is taken for the carbide to sink through the chute, during which process the acetylene escapes up the vent pipe.

When carbide in large pieces is used, or when the carbide is protected by being previously immersed in oil, the loss is small; but with the fine or dust-like material which constitutes several per cent. of each package of carbide the loss is greater than the available yield.

This fine carbide, even when made into a paste with kerosene oil, generates acetylene with almost explosive rapidity. The water is driven up the chute, the seal is often broken; and not only is the greater part of the intended charge of gas lost, but that in the bell also. In this event the lights go out, and the generator is temporarily disabled.

A smaller loss occurs by reason of the solubility of acetylene in the "sludge," which must be often removed from the generator. This is small, however —from one-half to one per cent. in usual practice.

If the "sludge" is allowed to become too thick with precipitated lime, even the large pieces of carbide sink with difficulty through the chute, and the loss of gas during the charging process is great.

In ease of management the dry machine, when it performs its functions properly, has some advantage. It is more quickly charged by those accustomed to it than is the wet machine.

On the other hand, the by-products of the dry machine are intensely disagreeable, as compared with the "sludge" of the wet generator, and ordinarily require handling more often.

The dry machine is certainly simpler in its best form than the wet, but neither is really complicated. All the pipes and connections are permanent fixtures, and each generator has only one or at most two cocks or valves.

To offset the devices whereby water is fed to the dry generator, the wet machine has some simple form of ratchet mechanism for dropping carbide into the water. Either is simple enough to be understood by those to whom the attendance of the apparatus is generally entrusted, and neither is likely, except through gross carelessness, to get out of order.

Before acetylene as an illuminant takes the place which it is destined eventually to occupy, material changes must be made in the method of its generation.

At the present writing, the problem does not

seem an easy one, nor does its solution seem immediately probable.

The gas-holder, with its inlet and exit pipes, its safety appliances and drainage system, is all that can be desired, and is, withal, extremely simple and durable.

In the author's opinion, the crux exists in accomplishing, without undue complication, three things.

The first is to introduce the carbide into the generator without loss of gas or admittance of air.

The second is to remove the lime under the same conditions ; and the third is to devise a means in the dry generator for the prevention of over-production and heating.

Until these desiderata are accomplished facts, the generation of acetylene will be at best a disagreeable performance, tolerated for the sake of the unrivalled light to be obtained from the gas.

In spite of its difficulties, the problem is not in the least hopeless, neither is there any insurmountable obstacle in the way of its accomplishment ; and it is to be fully expected that as the interest in acetylene becomes wider, as the demands for perfection in generators become imperative, more permanent apparatus will take the place of that which has temporarily served to tide the new illuminant over the period of its infancy.

IMPURITIES OF CARBIDE AND ACETY-LENE

In the manufacture of calcium carbide, in spite of the most careful effort, it is practically impossible to obtain either lime or coke which is entirely pure. For this reason the carbide, and, consequently, the acetylene generated therefrom, is alway contaminated with other products. The amount of impurities in the gas is generally so small as to be negligible. The principal impurities in the carbide are of importance from the fact that they are not acted upon by water, and when the carbide is converted into gas they remain within the generator either in the form of' lumps, like pieces of clinker, or of rounded nodules of various sizes. The nodules are of two kinds, one small and non-magnetic, which upon examination appears to be principally graphite, reduced from the coke by the intense heat of the furnace; the other nodules are hard, white, and glistening on the surface, sometimes showing signs of rust when exposed to dampness, generally possessing magnetic properties, and are of a hardness sufficient to scratch glass.

These nodules contain silicon and iron, with sometimes traces of aluminum and magnesia. They are somewhat similar to carborundum in their properties.

The contamination of acetylene, which results from the use of an impure carbide, was at one time considered a very serious element of danger. The only danger which could result from an admixture of foreign substances with acetylene, would be the possibility of forming in the mixture a self-igniting gas or compound of gases.

There are two gases — phosphoretted hydrogen and siliciuretted hydrogen — which are spontaneously inflammable upon contact with the air. Either of these may exist in acetylene produced from carbide which contained calcium phosphide or silicide.

The experiments, however, of Professor Vivian Lewes have shown that either of these gases would have to be present in a very large proportion in order to make an explosive mixture with the acetylene. He examined twelve samples of carbide from different sources in which the percentage of phosphoretted hydrogen ranged from a trace to nearly two and a half per cent., and found none under the most favorable conditions capable of producing a self-igniting gas. He discovered, also, that an actual admixture might be made containing as high as eighty per cent. of phosphoretted hydrogen introduced separately into acetylene without producing a self-igniting mixture.

It is believed that no authentic instance of accident from the self-ignition of acetylene has been recorded, and in the American carbide the proportion of calcic phosphide is, by careful selection of the materials, kept down to a minimum.

Nearly all acetylene which has not been carefully washed, shows the presence of a small quantity of ammonia together with some sulphuretted hydrogen.

The ammonia is detrimental in that it favors the production of acetylide of copper in the cocks of the burners, making them turn with difficulty after a time; but it is so easily removed by passing the gas through water, together with the sulphuretted hydrogen which accompanies it, that neither of these impurities need exist in the gas made in a well-managed generator.

It is said that much of the European carbide, manufactured upon the Continent, contains sufficient phosphoretted hydrogen to produce an irritating vapor of phosphoric acid in the rooms in which the gas is burned. In this country there has been no report of a similar trouble. As a matter of fact, the impurities which have been most often noted in the gas are due to overheating in the generator. When acetylene is subjected to a high temperature it splits up, as is well known, into certain polymers, such as benzine, styrolene, naphthalene, anthracene, tar and oily hydrocarbon. The decomposition is more complete as the temperature is higher, and is nearly in the order of the substances given above. At a red heat carbon is separated in the form of lampblack, and hydrogen is set free.

The objection to a gas containing impurities is

chiefly that of the money loss, owing to the poor
yield per pound of carbide.

The benzine vapor, hydrogen and other volatile
products, resulting from too high a temperature in
the generator, not only do not of themselves give
luminous flames when burned in an acetylene jet,
but they actually, by dilution, reduce the candle-
power of the remaining gas.

In addition to these disadvantages, the benzine
vapor is the frequent cause of the clogging of burners. It exists in the form of vapor condensed within
the pipes, and is carried to the burners by the outrush
of acetylene. There it is deposited upon the steatite
tip of the jet, and after being absorbed is carbonized
and causes a contraction in the jet orifice. Jets
which have once become smoky from this cause,
rarely recover their original power of illumination,
even if carefully cleaned. It is probable that a certain amount of the benzine, and other liquid which
has been absorbed by the steatite tip, remains there
and is gradually carbonized and driven into the orifice by successive heating of the burner.

The very general use of bicycle acetylene lamps,
which are made entirely of brass, has caused the
theory of the formation of acetylide of copper to
fall into disrepute. In these lamps, however, brass
is always used and not copper; and the lamps are,
as a rule, nickel plated, both inside and out. The
charges of carbide are frequently renewed, and the
lamp, by constant washing, is kept relatively clean.

The experiment of Bullier, who found that a polished copper plate was practically uninfluenced by acetylene for a considerable length of time, has contributed to the general feeling that the formation of acetylides is an exploded idea.

The author has experimented carefully upon this subject and is entirely satisfied, as a result of his work, that acetylide of copper is formed with great readiness whenever ammonia is present in the gas, in the presence of moisture. He furthermore finds that a polished copper or brass surface is much less affected than is one which has been already corroded and covered with oxide. He finds also that a surface fresh from the file or the turning lathe, covered with minute tool-marks, almost immediately shows the red coloration peculiar to acetylide of copper when subjected to the action of nascent acetylene.

It is not pretended that any real danger exists from this formation in the gas fixtures and pipes, but the gradual corrosion of the stop-cocks is annoying, and the author is convinced that it is a frequent cause of leakage. It is certainly inadvisable to use either copper or brass in the construction of the generator.

Another impurity almost always present in acetylene which has been recently produced, is a fine dust thrown off from the lime in the process of decomposition. This cloud of lime-dust is so light and its particles are so fine that it remains suspended in the gas for hours, and unless special care is taken to remove it by filtration through cotton or other

fabric, the dust enters the pipes and eventually clogs up the burners, and assists in cutting out the stop-cocks.

Another impurity which is of almost universal occurrence is air. In small proportions this has no other ill effect than to lower the candle-power of the gas, but it certainly should be excluded if, for no other reason than the fact that in certain proportions it makes an explosive mixture.

It will be understood that if the generator contains a certain quantity of air, this goes over into the gas-holder with the first gas which is generated, and there will be a time at the beginning of the gas-making when the mixture is explosive. The admission of air to the gas in any quantity is certainly inadvisable, and may be dangerous.

GENERATORS

In the earlier stages of experiment with acetylene, before suitable burners were devised, it was considered necessary to mingle a certain proportion of air with the gas, either in the generator or in a special mixer, before it reached the distributing pipes. Since this practice has been entirely given up, on account of its danger, it will be unnecessary to more than mention such apparatus as something to be carefully avoided.

The same may be said of such generators as deliver the gas under considerable pressure, and of the various machines for reducing it to a liquid form.

Generators for the production of acetylene gas under the very slight pressure necessary to send it to the burners may be divided broadly into three classes, each of which has many modifications, and in each of which the various operations have been performed in devious ways.

Some of the variants scarcely come under the classes into which they have arbitrarily been divided, but, as they are not numerous, it seems unnecessary, so long as they retain salient features of any one type, to classify them separately.

The three types of generator may be classed as follows:

1st. Generators which have the generator and gasometer separate, and in which the gas is produced by supplying water gradually and in measured quantity to a considerable portion of carbide contained in a closed vessel.

2d. Generators which contain both the carbide and water, with means for immersing and withdrawing the carbide successively by a relative movement of the carbide and water.

3d. Generators provided with means for dropping measured quantities of carbide into a large volume of water.

GENERATORS OF THE FIRST CLASS.

Dickerson Generator (Fig. 11).—In this generator, which was one of the first invented, a gasometer is connected by a pipe to a sealed chamber containing a layer of carbide and having within it a perforated tube communicating with a reservoir of water and closed by a measuring stop-cock. This cock is operated by the rise and fall of the gasometer bell. The plug of the cock is hollowed out and has two slots at right angles to each other. When one slot is in a vertical position, it is open to the water-reservoir. Water enters and fills the hollow of the plug. Upon the descent of the gasometer bell, the cock is rotated, the reservoir is shut off, and the second slot is brought into position to deliver the water in the hollow plug to the tube which conveys it to the carbide. As the gasometer rises, the parts resume their first position and the operation is repeated until the carbide is exhausted. In order to deliver the water quickly to the carbide, the axis of the cock is provided with a weighted lever, so that when the cock is turned past a certain point the weight falls, turning it the rest of the way.

Generator of Janson and Leroy (Fig. 12).— This generator has a gasometer which is provided with two retorts or generating chambers, either one of which may be put in action or cut out by means of the cocks r, r, C, C.

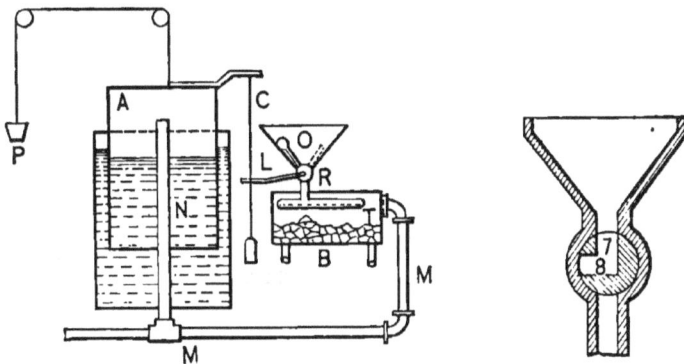

FIG. 11. Dickerson's Generator.

An elevated tank of water communicates with the retorts by means of a tube containing a cock (R) kept closed by a spring. When the bell of the gasometer falls, a finger (D) with which it is provided opens the cock (R) and delivers water to one of the retorts. When the carbide in that retort is exhausted, the further descent of the bell rings an electric alarm, warning one to so turn the cocks as to put the other retort into action. The first may then be disconnected, opened, cleaned of lime and recharged.

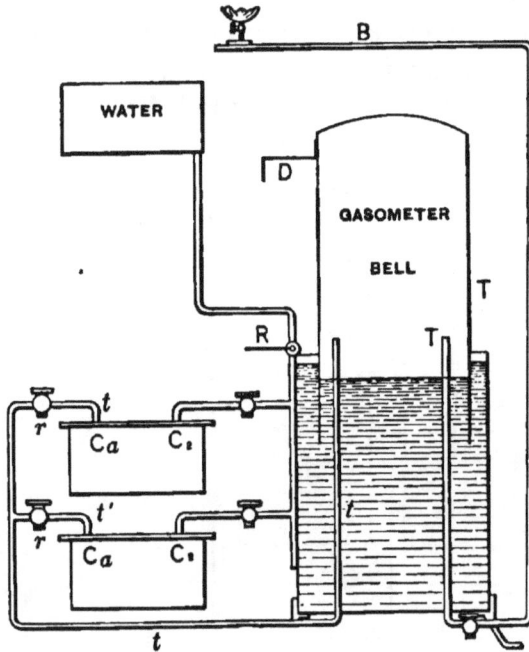

FIG. 12. Janson and Leroy Generator.

The Bon Generator (Fig. 13) has a cock (r′) operated by the rise and fall of the gasometer bell. A reservoir of water is placed on top of the gasometer and is provided with a water-gauge, so that, if a quantity of water is placed therein sufficient to decompose the carbide placed in the generator, an inspection of the gauge will show how nearly the charge is exhausted. The water descends by the

tube G and falls into a funnel (G') which conducts
it by an inverted siphon to the carbide.

The carbide is placed in a pan (F) which is divided
into compartments, numbered from 1 to 12 in the

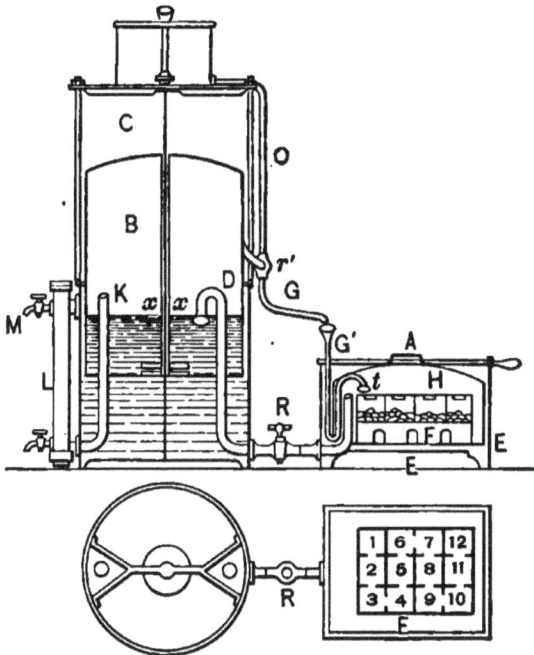

FIG. 13. Bon Generator.

lower figure. Each compartment contains such a
weight of carbide that the gas which it disengages
will just fill the bell of the gasometer. If too great
a quantity of water enters, it attacks the carbide in
the first compartment only, so that there is no over-
production of gas. Each compartment communi-

cates with the one before and the one following by means of a notch cut in one of its walls.

The water falls always into the first compartment, and, after the carbide it contains is decomposed, overflows into the next, and so on until all are successively filled. The pan containing the carbide is placed in an outer case containing water, which keeps it cool and acts as a seal in conjunction with the bell (H) held in place by a bar (A).

The gas enters the holder by a curved pipe (D) below the water line; it is washed by bubbling up through the water. A dryer (L) in the delivery-pipe may be filled with any absorbent of moisture or, as is sometimes the case, with lumps of carbide, which not only absorb the moisture but give off additional gas.

The Souriou Generator (Fig. 14).—This is provided at its base with a circle of retorts each containing sufficient carbide to generate gas enough to fill the gasometer.

Each retort is provided with a weighted lever controlling its water-supply, and each lever is normally held in position by a latch to close the cock which it operates. The latches are successively opened by the motion of the gasometer bell.

The Clausolles Generator. This has a single retort surmounted by a reservoir of water. A cock connecting the two is operated by the gasometer bell as it rises and falls. In this machine it is intended that the admission of water shall be proportional to the gas used.

The only novelty is a tube open at both ends, which projects from the top of the bell and extends downward to nearly its entire depth. In case of

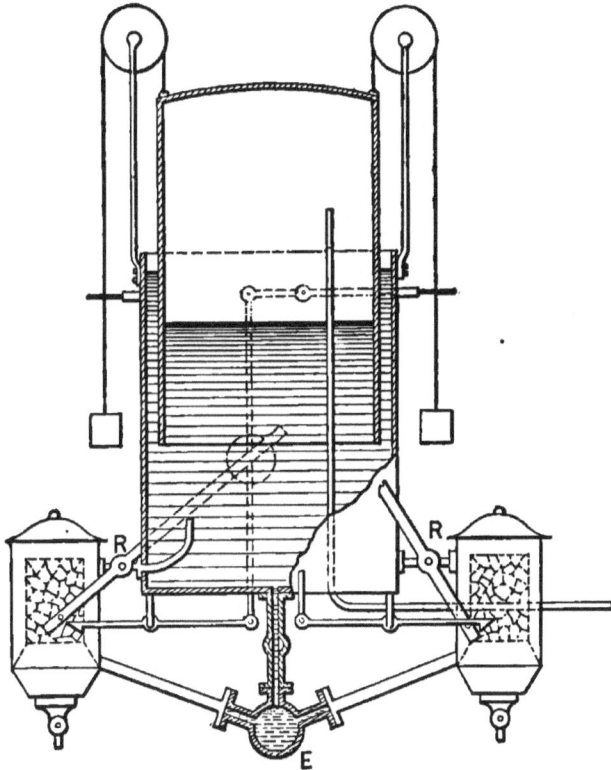

FIG. 14. Souriou Generator.

over-production, the lower end of the tube is lifted out of the water before the bell is quite full, allowing the surplus gas to escape. By this means the spattering of the water, which attends the escape of gas from the under edge of the bell itself, is avoided.

The Voigt Generator (Fig. 15) differs from the last only in having the controlling cock actuated by a rack and gear-wheels, and in passing the gas through a cooling coil on its way to the holder.

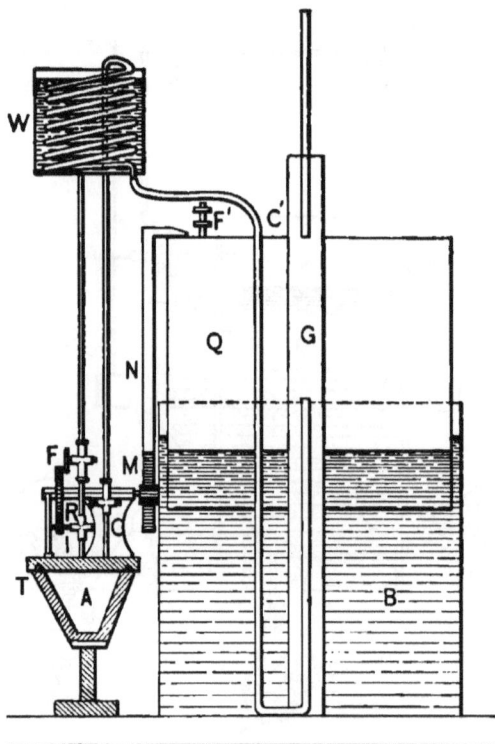

FIG. 15.   Voigt Generator.

The Humilly Generator (Fig. 16) is an ingenious arrangement of this type of machine. The gasometer in this machine is small and acts principally as a regulator. The generator consists of a tightly closed bell weighted with a lead disc (p).

Within is a basket of carbide, through which a tube
(C) projects and is surmounted by a conical distributor (I). The retort is immersed in the water contained in the outer vessel by which it is kept cool.
Some of the water finds its way up to the tube C,

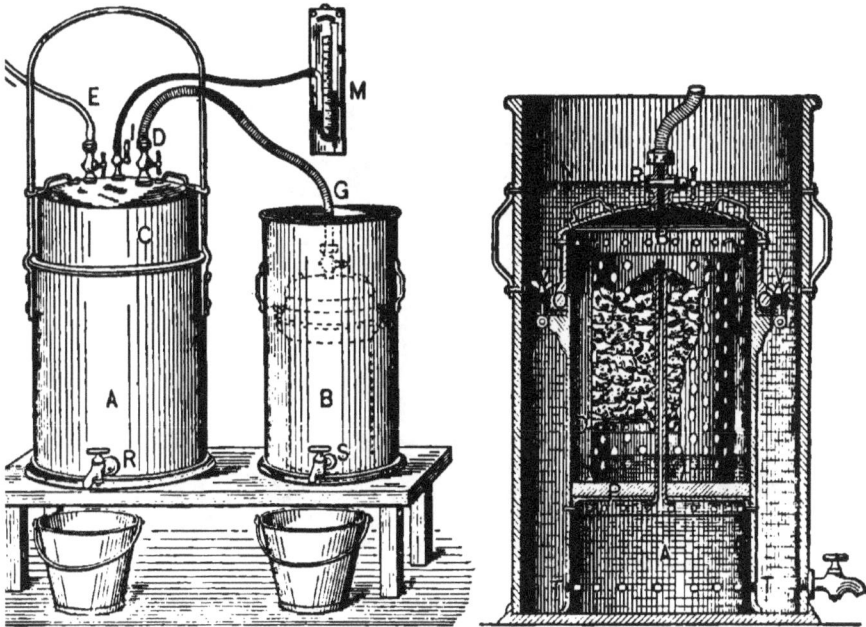

FIG. 16. Humilly Generator.

whence it runs down the inclined surface of the cone
I and drips upon the carbide. As the pressure
increases the flow of water is arrested.

An improved form of this generator is shown in
Fig. 17, in which the carbide is divided up among a
number of pots having holes through their sides at

different levels. By this device the water enters one pot after another, so that only a small amount of carbide is acted upon at a time.

FIG. 17.   Humilly Generator.

The Exley Generator (Fig. 18), which has been much used, is on a similar principle, but has the carbide contained in two retorts (R, R') fastened to the side of the water-vessel A, with which they communicate by means of tubes (T, T).

In action, the cock v is closed, while v^1, v^2, v^3 are open. The water coming through the hole O passes by the pipe T to R', where it attacks the car-

bide. The gas generated passes by way of the pipes
D′, D to the upper part of the reservoir (A), where
it displaces the water, forcing it into B by way of
the drop-tube P. As gas accumulates in A, the
level of the water sinks until the hole O is reached,

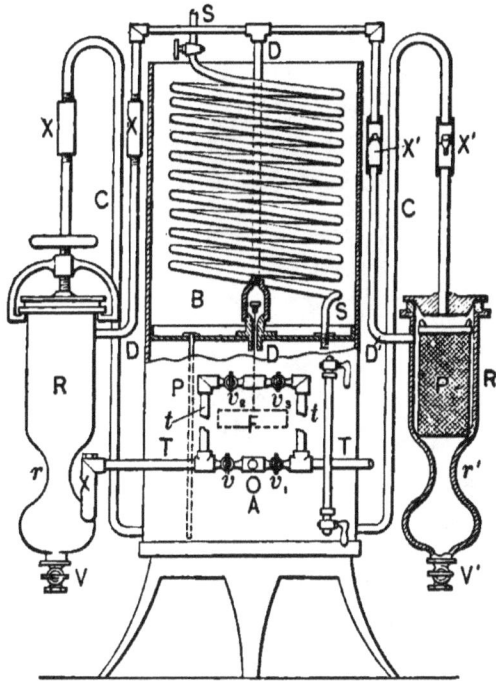

FIG. 18.   Exley Generator.

when no more can enter the retort.   The gas is led
to the burners through the cooling-coil (S).   When
the contents of retort R′ is exhausted, the water in
A rises, as gas is used, until it flows through the
branch pipe and open cock (v^2) to the retort (R),
where the same action takes place as before.

When this happens, which may be determined by the reading of the water-glass attached to the apparatus, the first retort is opened and recharged. During this process the cocks v^1, v^3 are closed.

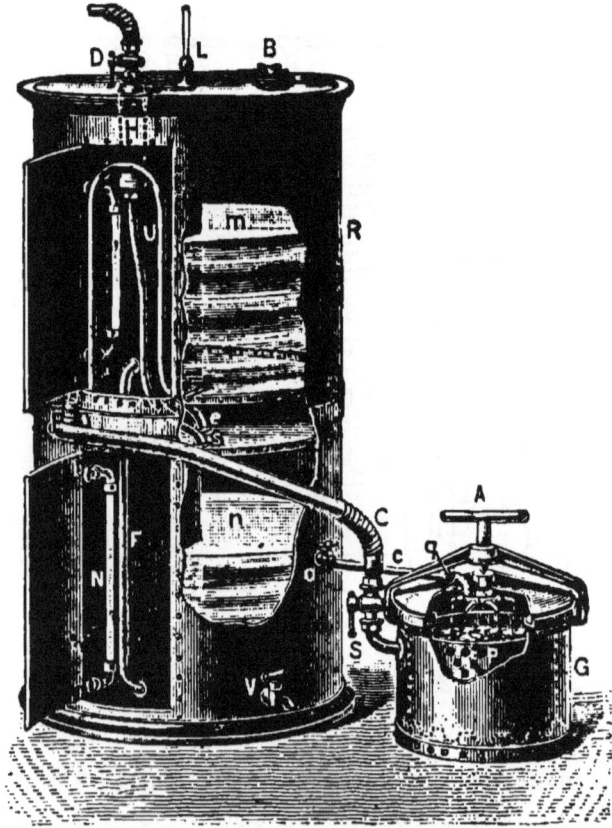

FIG. 19.   Gillet & Forest Generator.

Safety-pipes having valves (X) to prevent return of the gas to the retorts are provided.   There is considerable unnecessary complication about this generator, and, as the level of the liquid determines the pressure of the gas, a fluctuating delivery results.

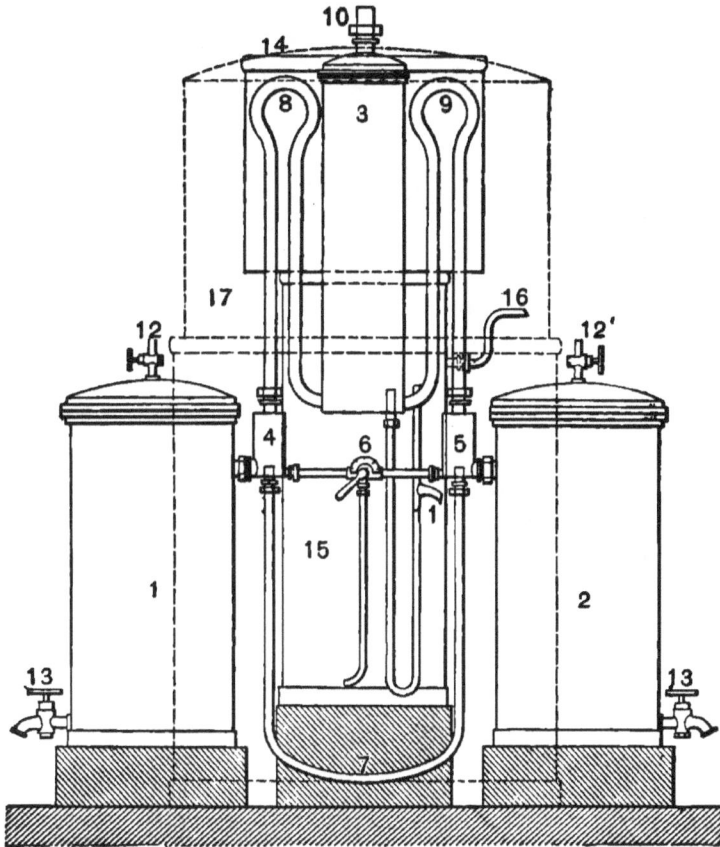

FIG. 20. Deroy Generator.

The Gillet & Forest Generator (Fig. 19) operates on the same principle of displacing the level of water contained in the gas-holder. The retort

(G) contains a basket of carbide (P), and is closed by a cover and screw (A). Water enters by the small tube c, while gas leaves by tube C. As gas accumulates in the holder, the water is forced into the upper section. When the level of the water in the lower part reaches the mouth (o) of the tube (c), the flow ceases and a fairly close regulation is for a time effected.

The Deroy Generator (Fig. 20) is said to give a very regular output of gas. It consists of a gasometer (15) with bell (17), to which are connected two retorts (1, 2) by curved pipes (8, 9), which convey the gas first through a washer (3).

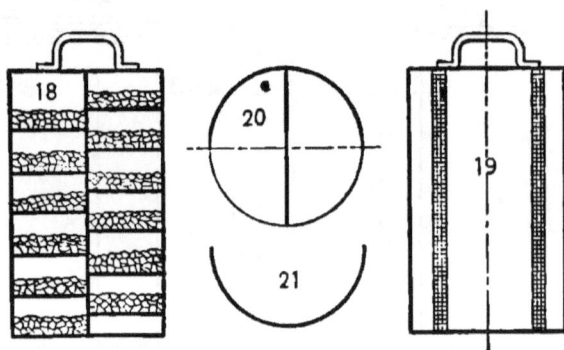

FIG. 20A.

Water from a reservoir, fed from a second reservoir (14), enters retort 1 by way of the cock 6. The retorts contain carbide subdivided into separate cells (Fig. 20A) by a series of discs and rings, so that the different portions are at various levels. The water, as it rises in the retort, can reach only one cell at a time, and it is intended that the contents of each cell shall, when decomposed, fill the bell with gas.

The supply of water to the carbide is controlled by the pressure in the retort and the level of the liquid in the reservoir, which is provided with an overflow (16). When retort 1 is exhausted, the

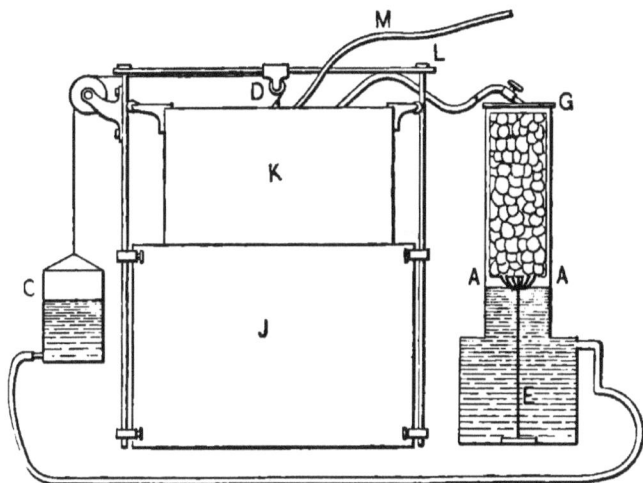

FIG. 21.   Chesnay & Pillion Generator.

water rises until it flows by the curved tube 7 to retort 2.   No. 1 may then be recharged.

Before that is done, the cock 6 is turned to supply directly retort 2.

The Generator of Chesnay & Pillion has the usual arrangement (Fig. 21) of retort and gasometer.

The novelty which makes it worthy of notice is that the reservoir of water is raised and lowered by the bell, being suspended therefrom by a cord and connected to the retort by means of a rubber tube. T. O'CONNOR SLOANE, of New York, has made a very pretty application of the same idea for lantern projections. It may be here remarked that the acetylene light answers admirably for the magic lantern, and, with a small generator of this type, is convenient and cheap.

In **Ragot's Generator** a very ingenious arrangement is shown.

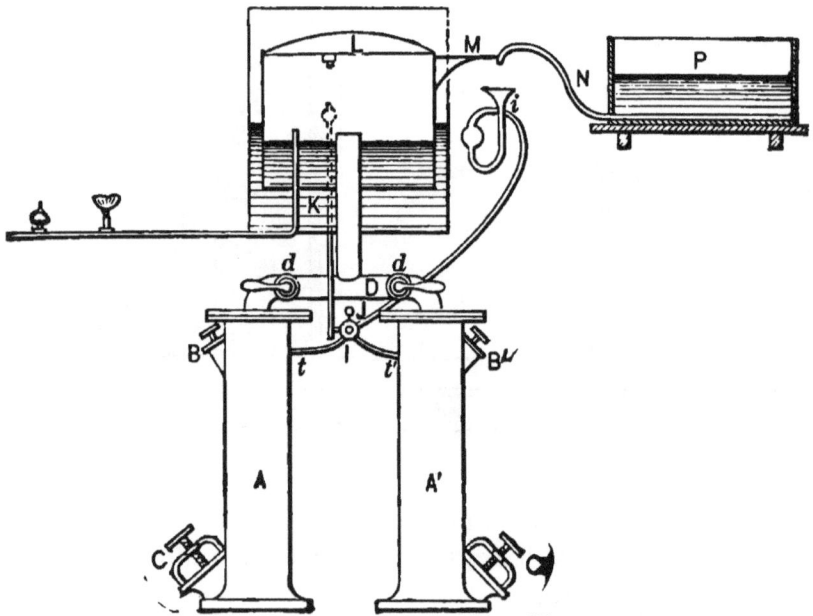

FIG. 22.   Ragot's Generator.

In order to avoid the varying weight of the full and empty reservoir of the last machine, he supports the reservoir of water and moves the flexible outlet leading from it by the rise and fall of the bell. This

is illustrated by Fig. 22, which also shows the pair of
retorts used and the means for putting either one
into operation alternately.

La Phare Generator.—In this generator (Fig.
23) an elevated .water-supply (E) is connected by a

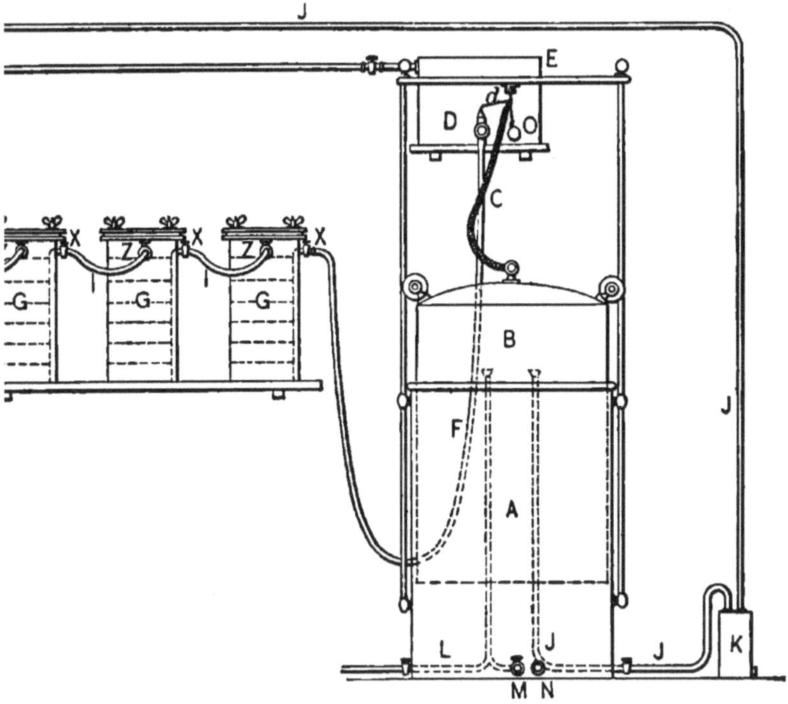

FIG. 23.   La Phare Generator.

flexible tube to the first of a series of retorts (G, G)
containing carbide in superimposed pans.   Each
retort is connected by a rubber tube to the next in
series, and the last by way of a wash-bottle (K) to the
gasometer.   A cock (D, D), normally held closed by a

spring, is opened by a cord (C), which is pulled when the gasometer bell sinks to a certain point. Water enters the first retort, from which the gas passes by way of the opening Z through the other retorts to the gasometer. When the first retort is exhausted, the water overflows into the second, and so on. The first retort is then disconnected from the series and the water-tube connected to the second. A water-supply pipe and float-valve keep the reservoir D always full.

The Springfield Generator.—This machine, which was perhaps the first to recognize the desirability of storing acetylene in an underground gasholder, is shown in Fig. 24.

This generator is of the intermittent type, requiring to be recharged with carbide each time the gasholder is emptied. Owing to the unequal pressure upon the gas in the holder, corresponding to different water levels, a pressure-regulator is required in the main delivery-pipe.

A and A' are iron tanks, each having a capacity sufficient to hold the gas generated by one charge of the generator.

B, the generator, is a cast-iron cylinder.

C, the lid or cover of the generator.

E is a gallery supplied with mercury, in which the lip of the lid is immersed when the generator is closed, in order to make a gas-tight seal.

F is a clamp for holding the lid shut, and forms the lever of the gas-cock (D). If desired, this may be secured with a padlock.

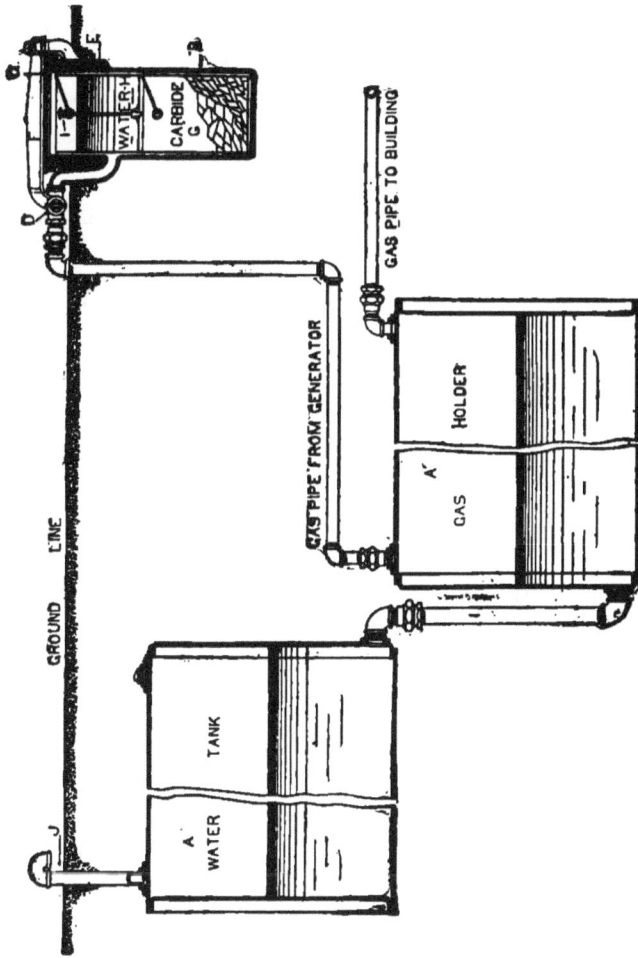

FIG. 24. Springfield Generator.

G is a galvanized iron pail or bucket, containing the carbide of calcium.

H is a similar bucket, containing water.

I is a handle, controlling a valve in the bottom of the water-bucket and reaching the top of the bucket.

The return bend (J) is screwed off and water is poured in until the lower tank (A') is full, leaving the upper tank (A) nearly or quite empty. This supply of water is permanent. This introduction of water will be accomplished best when the lid (C) of the generator is removed and the clamp (F) lies horizontal, which will furnish a vent for the air in tank A'.

The pail G, partially filled with carbide of calcium, is placed in the generator first. The handle I being turned, the water in the pail H will run slowly down into the pail containing the carbide, and gas will immediately begin to generate. The cover C should be placed in position and fastened with the clamp F immediately after opening the valve I.

The gas formed by the decomposition of the water in contact with the carbide will accumulate on the surface of the water in the tank A', and the water will all be forced up into the tank A. The tank A' will be full of gas, and the tank A full of water.

The gas is now ready for use.

A pressure-gauge is located inside the building lighted, at any convenient point from which may be determined by a glance, at any moment, how much gas remains in the machine. A pressure-regulator located within the building lighted insures the delivery of the gas to the burners at a proper pressure.

The Napheys Generator.—In the upper section of the steel tank is suspended a cylindrical steel

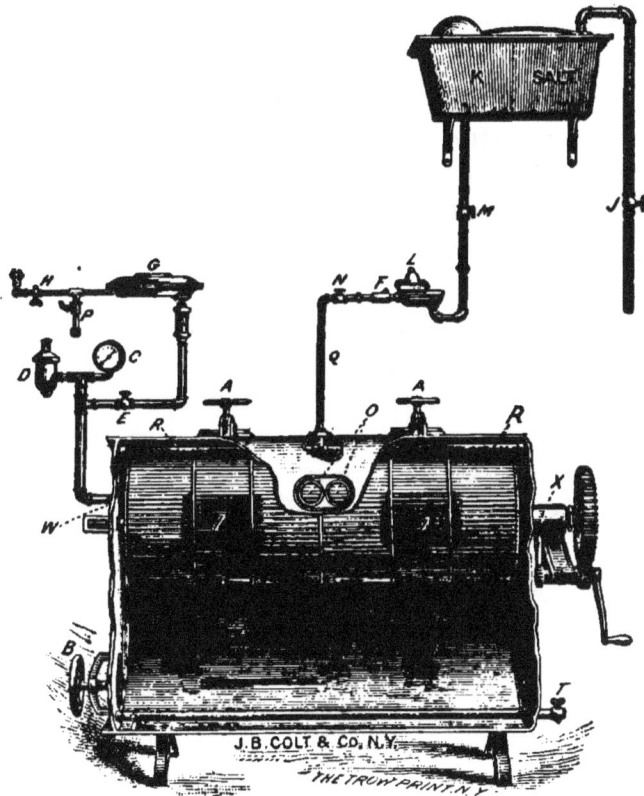

FIG. 25. Napheys Generator.

cage. The cage is supported by a shaft which passes through a stuffing-box (X) and has attached to it a large gear-wheel. This in turn is geared to a crank, and turning the latter slowly revolves the cage. Carbide is placed into the cage through ports A, A, and the slide doors I, I, which are then slid back into place, and then the ports A, A clamped, so as to be air-tight.

The auxiliary reservoir K is supplied with water from any available source, either the city supply or house tank. Salt is placed therein to prevent freezing, if the water is exposed during cold weather. Water falling by gravity passes through cock M, automatic-regulator L, check-valve F, cock N, and pipe G to sprinkler-pipe R, R. The latter should be perfectly level, so as to spray all parts of the carbide evenly. As the water comes in contact with the carbide it is rapidly decomposed and acetylene generated. Cock T is then opened to allow the air contained in the generator to escape. This accomplished, cock T must be closed. The acetylene finds exit through pipe W, passing through valve E, and the gas-pressure regulator G to H, where house connection is made. It also operates the indicator C, and, if pressure should for any reason become too high, would escape through the safety-valve D. As the pressure rises in the tank it correspondingly rises in water-pipe Q and the automatic-regulator L. When a certain pressure is reached, the regulator L automatically cuts off the water-supply, preventing any further admission of water upon the carbide.

Each day the cage should be turned over. This allows the residuum (slaked lime) which has accumulated on top to fall through the cage to the bottom of the generator. From there it is withdrawn through port B.

The openings in the safety-valve and mercury blow-off are to be piped out-doors. A whistle may be attached to the end of the pipes if desired, so that in case either are brought into operation instant notice will be given and the disarrangement stopped.

The Wallace Generator (Fig. 26) has been used to a limited extent in this country.

It consists of a cast-iron retort—shown at the left of the cut—closed by a door bearing upon a rubber gasket, a gasometer, water-supply kept at constant level by a ball-float valve, a cooling-coil within the reservoir through which the gas flows on its way to the holder, and a pair of water-seal safety devices, one below the retort and one at the bottom of the gasometer.

The generator is charged by placing in the retort a pan containing about fifteen pounds of carbide. The door is closed and the cock leading to the gasometer through the cooling-coil is opened.

The water-valve is then opened, allowing water to spray upon the carbide.

A heavy, annular weight suspended over the gasometer is lifted by the rising bell at a certain point in its travel. When this happens, the pressure of gas in the retort is somewhat increased, and, as the level of water in the reservoir is very slightly above the top of the retort, its flow is arrested.

A pressure-gauge is attached to the machine, which

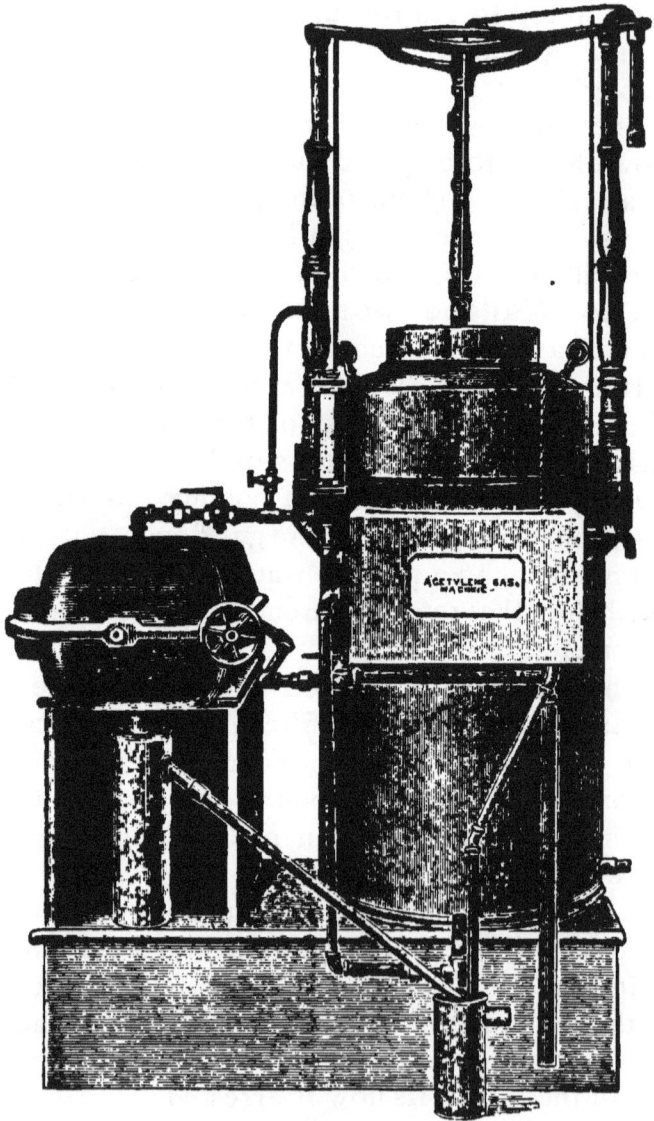

FIG. 26.   Wallace Generator.

is also safeguarded against excessive pressure by the
water-seals contained in the two safety-cans.

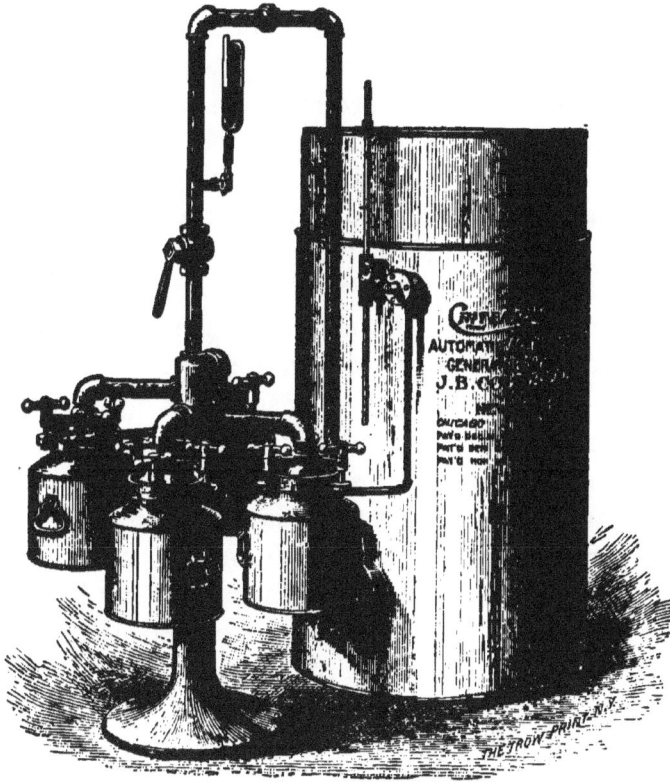

FIG. 27. The "Criterion" Generator.

The " Criterion " Generator (Fig. 27).—This
generator consists of a stand for carrying the carbide-holders and a gasometer for regulating the
supply of water, keeping the gas pressure constant
and taking care of the surplus gas. A condenser

for cooling the gas is at the base of the gasometer. The water-regulator is placed on the side of the gasometer, from which the water supply is taken, except in large generators, in which a separate water tank is used.

The carbide-holders are made in various sizes and are attached to the stand by pipes which radiate from a central upright, but each holder is on a different level from the others; four, eight, or twelve of these holders, according to size, can be conveniently arranged in sets of four, one set above the other, the gasometer being of the same capacity for any number of holders of a given size.

As the gas is used, the gasometer descends, the water-regulator is opened and water is admitted to the stand and flows into one of the carbide-holders. In the small generators the gas passes into the stand by the same pipe through which the water enters the holder, then into the gasometer and to the burners. If an excess of gas is formed, the gasometer rises and the supply of water is automatically shut off.

When the carbide in one of the holders is exhausted, the water rises into the next holder.

In case of too sudden generation of gas the water instantly stops flowing into the holder, because the rush of gas over the water in the same pipe holds the water back and prevents the formation of gas until wanted.

SECOND CLASS.

Generator of Allemans & Stemmer (Fig.
28).—A funnel-shaped vessel (G), having a gas-tight
cover, depends into a closed case (T). The carbide is

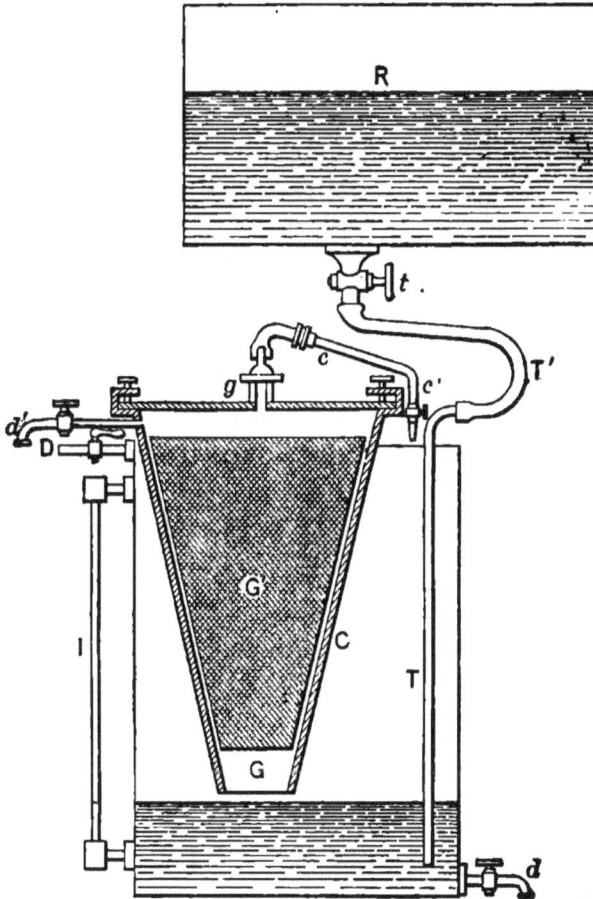

FIG. 28. Generator of Allemans & Stemmer.

inserted in a conical basket. Water from an elevated reservoir enters T until the carbide is reached, when the pressure of the gas generated regulates the supply. The conical mass of carbide is intended to equalize the production of gas by bringing a larger surface into action as the carbide is more and more exhausted.

d'Arsonval has designed a remarkably good generator for experimental use. It consists (Fig. 29) of an outer case containing a bell. Within the bell is hung a basket of carbide, introduced through a water-sealed cover. The water in the gasometer is covered with a layer of oil. When the bell sinks, the carbide suspended from it enters the water, giving off gas, and, under a constant rate of use, it soon finds a position in which it remains stationary, the production being then just sufficient to supply the demand.

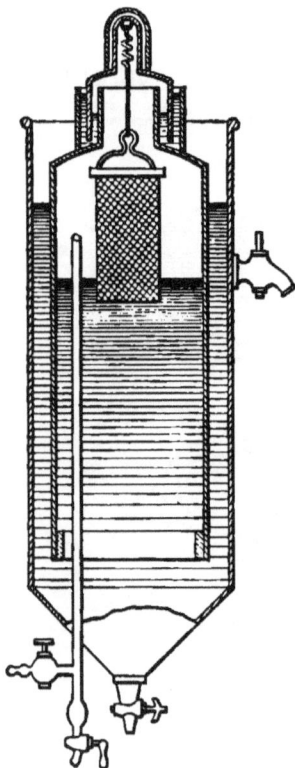

FIG. 29. d'Arsonval Generator.

When gas is no longer used, the production continues for a short time, the bell is lifted, and with it the carbide. The oil, as the carbide passes through it, displaces the water which had previously been

absorbed, preventing further action.   When the
carbide is lifted entirely out of the liquid, the layer
of oil, by preventing evaporation of the water, keeps
the moisture from slowly producing gas.   Generators on the same principle. without the layer of oil,

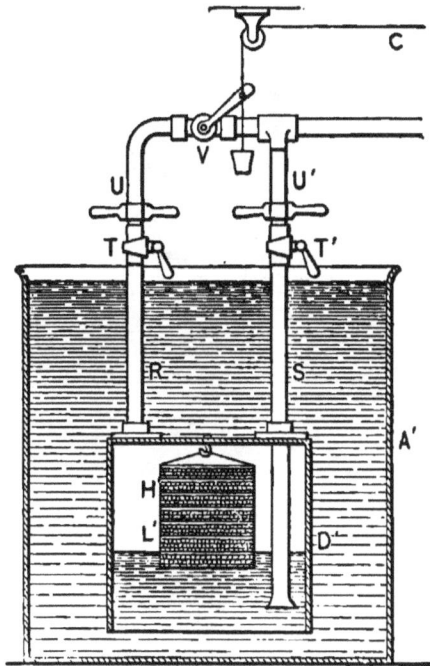

FIG. 30.   Gabe Generator.

are much used for lantern-work, but the gradual
evolution of gas when not in use is a serious draw-
back and is entirely overcome by this simple expedient.

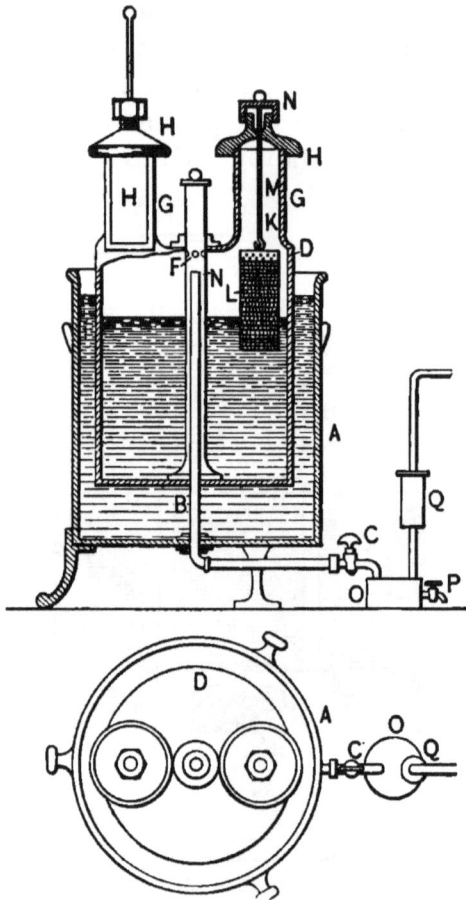

FIG. 31.   Gabe Generator.

The Gabe Generator exists in two forms: The industrial model (Fig. 30) has a basket of carbide hung in a bell contained in a large outer case filled with water.   The operation is obvious.   The gas goes through a washer to a gasometer, or is taken directly to the mains. In the smaller model (Fig. 31), the carbide is contained in a number of baskets in the top of a rising and falling bell. One after another is pushed down by means of the rod from which it is suspended until it enters the water. When all are exhausted, the generator must be put out of service while being recharged. The simplicity and cheapness of this type of machines recommend them to experimenters,.but the fact that the bell must be opened and any gas remaining therein lost before recharging makes them unfit for practical application on an extended scale.

THIRD CLASS.

The carbide chute machine seems to offer the best solution of the problems involved in making a continuously acting generator, because in it the carbide may be added and the lime removed without admitting air or interrupting the production of gas. Moreover, since there is always a relatively large quantity of water surrounding the carbide, no heating takes place.

Some of the earlier generators of this class required the use of granulated or powdered carbide, and although this form is not now in vogue, these machines are interesting. They will be first described.

The Marechal Generator (Fig. 32) consists of a closed case (B) on which is mounted the hopper (A) filled with powdered carbide. A sort of stopcock (R), having a pocket in its circumference, is ro-

tated by the movement of the piston (P), which the
fluctuating pressure of the gas causes.   At each

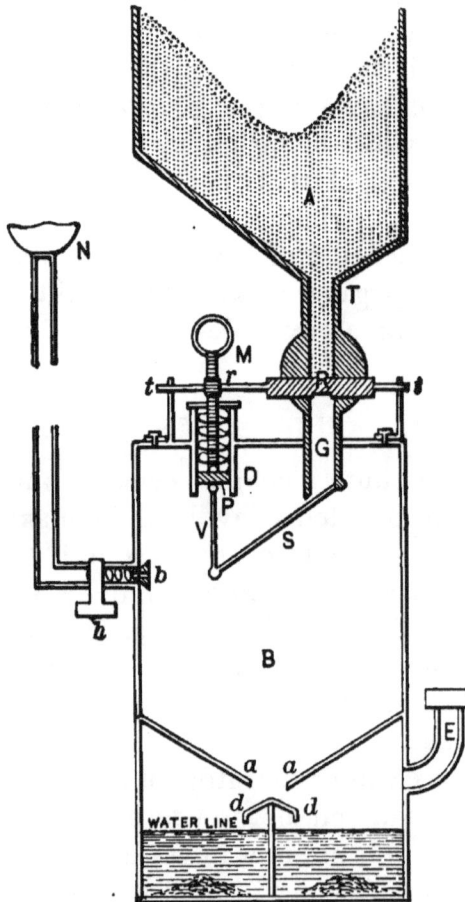

FIG. 32.   Marechal Generator.

movement a small quantity of carbide is delivered
to the tube (G) and by the movement of the link (S)

is allowed to fall into the water at the base of the machine.

The Thivert Generator (Fig. 33) has a hopper (A) and reservoir (C) carried by the bell of a gasom-

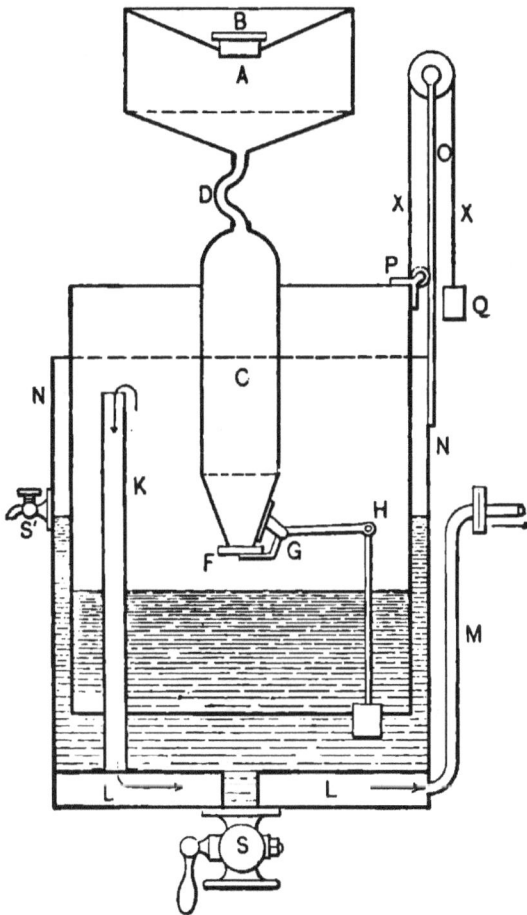

FIG. 33. Thivert Generator.

eter. The mouth of the reservoir is kept closed by means of a disc (F) and weighted lever (H).

Upon the descent of the bell, the weight rests on the bottom, raising the lever and depositing a small amount of the pulverized carbide in the water. As soon as the bell rises, the cover (F) is closed by the weight. Deposited lime is drawn off through the large cock S.

The Bouneau Generator (Fig. 34) is on the same plan, except that the weight is attached directly to a conical plug which closes the mouth of the carbide receiver. When the bell sinks the plug is lifted, letting carbide fall into the water.

Leroy & Janson have placed the hopper and valve within the gasometer in order to avoid the varying weight on the bell which results from carrying the carbide on that member of the apparatus (Fig. 35). As the bell descends, a rod (L) is pressed, opening the valve (M), which allows carbide to issue. The gas passes by tube (D) to a drier (E) filled partly with pumice-stone and having a layer of calcium carbide, in lumps, in the upper part.

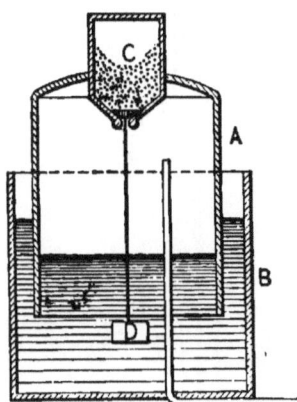

FIG. 34. Bouneau Generator.

The Lequeux Generator is made in several ways. A simple form for industrial use is shown in Fig. 36. One or more inclined tubes communicating through a water-seal with a gasometer are divided by a central partition extending from the top down-ward to a point well below the level of the water contained in each. Carbide is dropped by hand through an opening on the lower side of the parti-

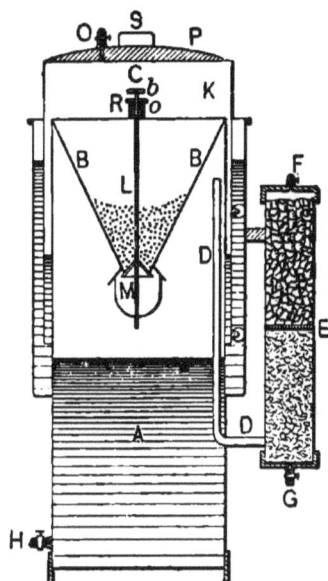

FIG. 35. Leroy & Janson Generator.

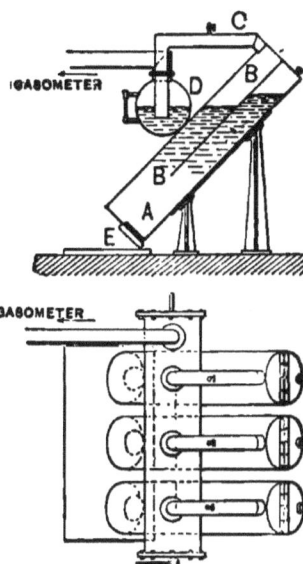

FIG. 36. Lequeux Gener-ator.

tion at the top of the tube. It falls to the bottom, and the gas which is disengaged, rising vertically, enters the part above the partition B, from which it issues by tube C.

Waste lime is removed through the opening E.

Another form is shown in Fig. 37, in which the chute K projects from the side of the generator A. A cover, closed by a water-seal, closes the generator,

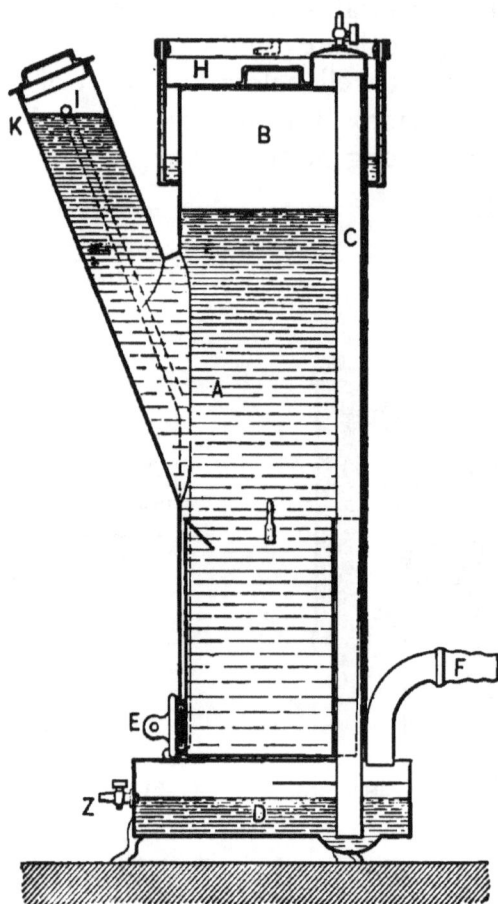

FIG. 37. Lequeux Generator.

which contains, at the bottom, a bucket for removing the lime bodily. Carbide in lumps is dropped
down the chute K into the bucket. Gas accumulates
in B while the water is displaced into the chute.
C is the delivery-pipe, dipping below the surface of

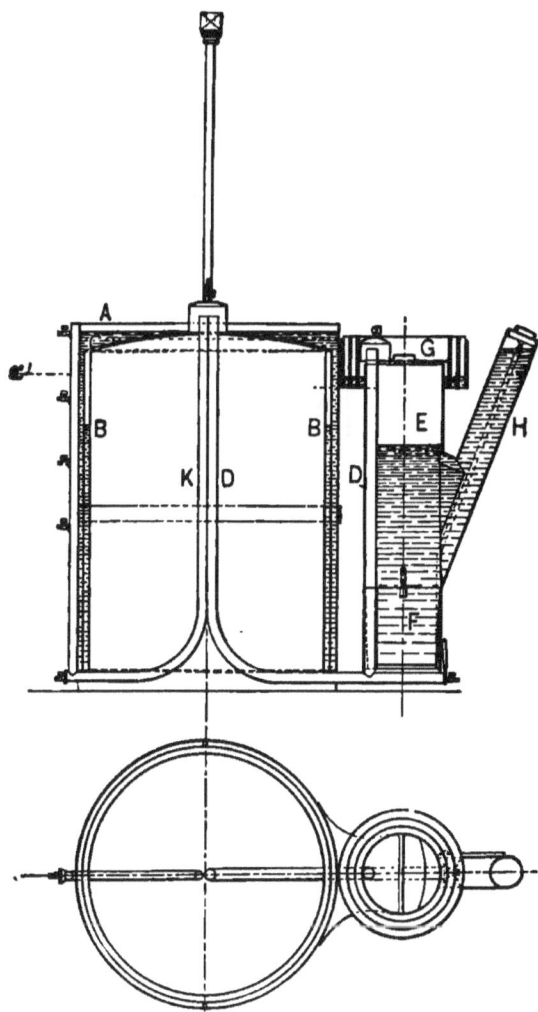

FIG. 38. Lequeux Generator.

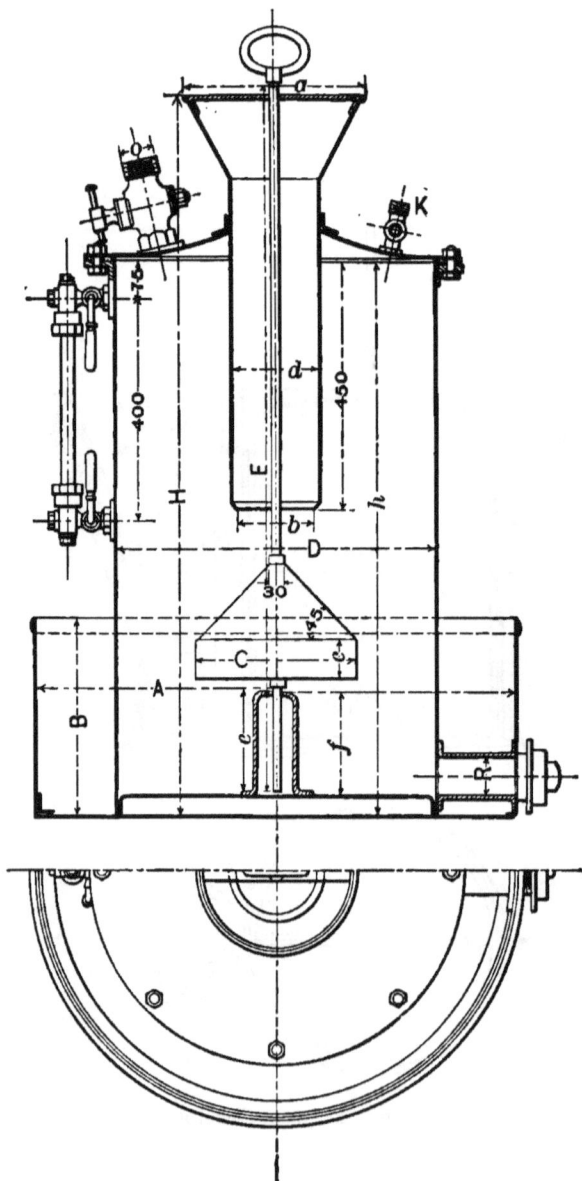

FIG. 39.   Société le Gaz Acetylene.

water in a combined seal and washer contained in the base of the generator. The exit-pipe F may be joined to the burners by a rubber tube. This generator, which is one of the best, is shown in combination with a gasometer in Fig. 38. In this arrangement the water-seal (G) is multiple.

La Société le Gaz Acetylene has built a larger and more important generator, shown in Fig. 39, to be used in conjunction with a gasometer. The drawing is self-explanatory. A central tube (E) answers as a chute, below which there is placed a conical deflector (C). The generator is filled to the line D with water. The vessel B, in which the generator rests, is filled with water. The lime is drawn off through R.

The Patin Generator (Fig. 40) differs from the Lequeux only in having its chute filled with oil, which floats on

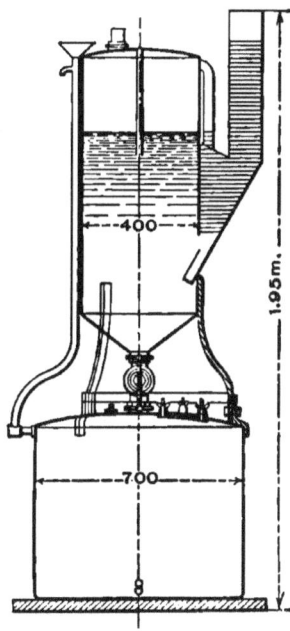

FIG. 40.   Patin Generator.

the water and delays the disengagement of gas from the carbide dropped into it. The plan is of doubtful advantage, as the carbide is constantly carrying oil into the generator, where it is broken up and changed into a disagreeable scum, which is difficult to remove and dirty to handle. This generator is

conical at its bottom, where it communicates, by
means of a large cock, with a lower vessel filled
with water, into which the larger part of the lime

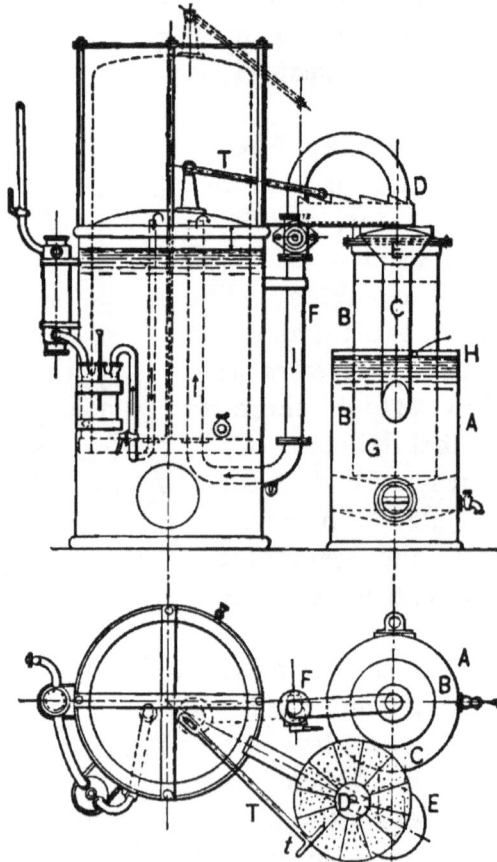

FIG. 41.   Seguin and de Perrodil Generator.

drops.   When sufficient deposit has accumulated,
the cock is closed and the lime removed from the
settling-tank.

The Seguin and de Perrodil Generator
consists (Fig. 42) of a pair of concentric cylindrical
vessels (A and B) filled with water. A chute (C) provided with a funnel (E) permits the introduction of
the carbide, which falls into the basket G. The car-

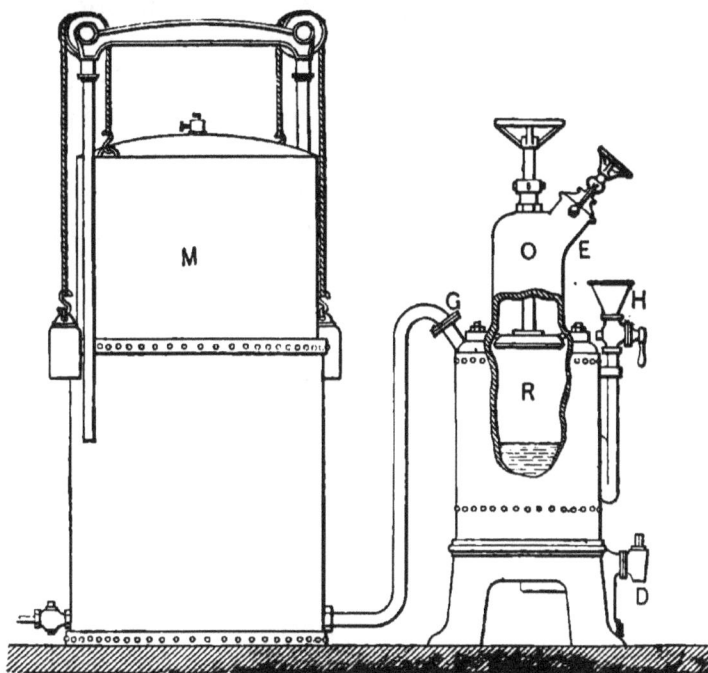

FIG. 42. Bertrand-Taillet Generator.

bide is placed in a distributor (D) turning on an axis
and provided with a ratchet-and-pawl mechanism
operated by the fall of the gasometer bell. Each
time the bell descends, the lever T engages with a
tooth of the ratchet D, whereby the wheel is rotated
until one of its compartments is brought over the

chute and the contents dropped into the generator. This is a truly continuous machine in every sense. It suffices to keep the distributor pockets filled with carbide, and to occasionally remove the deposit of lime, to keep it in operation indefinitely.

The Bertrand-Taillet Generator is made in two forms. One requires a separate gasometer of considerable size, as shown in Fig. 42.

A vessel (R) is surmounted by a dome (O) having a tubulure (E) for the introduction of carbide and a screw-operated valve closing its mouth. The carbide is emptied wholly or in part into the water by manœuvring the valve-wheel.

The other form is shown in Fig. 43. It consists of a gasometer having in its bell a number of pockets closed on the outside by tight covers, and by hinged bottoms within the bell. The bottoms are held shut by weights hanging from chains of various lengths, so that the contents of the pockets are deposited in the water successively by the repeated descents of the bell.

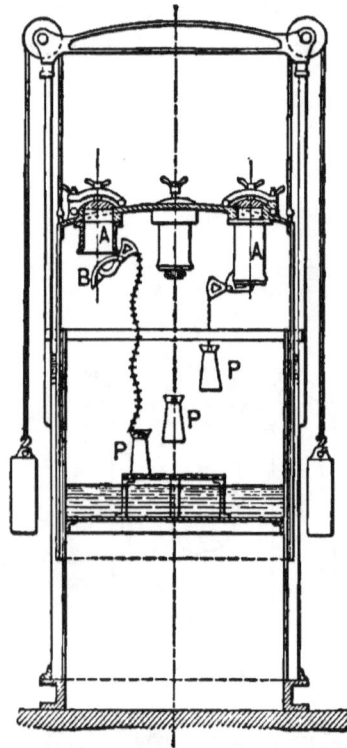

FIG. 43.
Bertrand-Taillet Generator.

ACETYLENE LAMPS

MANY forms of portable acetylene lamps have been designed, but none, thus far, has been very success-ful. While the light is in every way desirable, there is some fear of the lamp itself existing in most minds, and even if this is overcome it must be ad-mitted that the best acetylene lamps are troublesome to manage. They require a greater degree of intel-ligence for their successful handling than do the or-dinary oil-lamps, and, in consequence, it is scarcely advisable to trust servants to keep them in operative condition. In addition to these drawbacks to their use, there is the unavoidable odor from the spent carbide, which diffuses itself throughout the room in which the lamp is recharged. In spite of these objections, there are lamps on the market which are most attractive in appearance and for which the manufacturers claim immunity from danger or odor.

In a well-made acetylene lamp the danger from fire or explosion should be very slight, for the entire amount of carbide used for one charge does not usually weigh more than a few ounces. This, if the gas from it was all generated at once, could produce not more than three or four cubic feet; and, as the rate of generation is at a maximum rarely more than enough to supply one burner, the chance for dangerous leakage is small.

Again, in case of explosion of the gas in the lamp, the shock would scatter the carbide and water harmlessly about, whereas the explosion of an oil-lamp is generally followed by a shower of burning oil.

Lamp of Decretet and Le-jeune (Fig. 44).—The carbide is placed in a closed generator (G) in a perforated basket (S). The generator is placed in the body of the lamp, which is filled with water. The water, which serves both for cooling and for generating the gas from the carbide, enters the generator by the tube D, rises to the top of the carbide and drips thereon. As pressure accumulates, the water-supply is arrested and the valve at the top of tube D closes by its own weight. *Re* is a safety-valve. Two funnel-shaped plates (*ch*) are supported above the carbide-holder, which receive the moisture carried off by the gas, and from which it drips back upon the carbide.

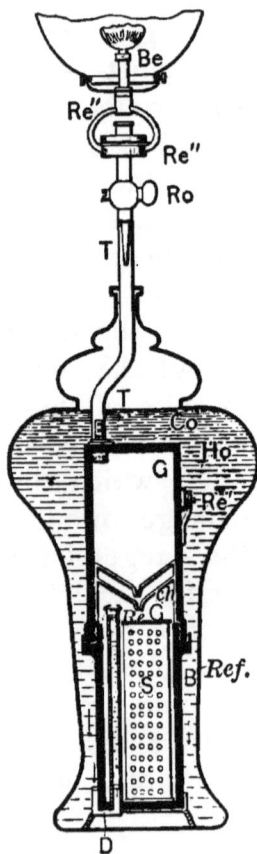

FIG. 44. Decretet and Lejeune Lamp.

by the gas, and from which it drips back upon the carbide.

Lamp of Gossart and Chevallier (Fig. 45).—
This is one of the simplest and best lamps yet designed. The water which is contained in the upper
chamber falls drop by drop through the bent capillary
tubes upon a mass of carbide placed in the bottom of
the lamp. The amount of water is so small that the
lime is reduced in the form of a dry powder. About
five ounces of carbide forms the charge. In case of
increased pressure the flow of water
is arrested, and in the event of an
abnormal increase the gas would
pass out and escape by way of the
capillary tubes. It is said that,
in full operation, the temperature
of the carbide never rises above
125° F., and that the water-supply
is so sensitive that, as the flame is
turned down, the drops fall less and
less frequently until they stop, when
the flame is turned out. It is also
claimed that the production of gas
stops at the same moment. There
are no valves, cocks, or other complications. Nothing more simple
could be desired.

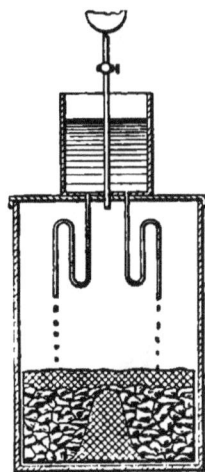

FIG. 45. Gossart and
Chevallier Lamp.

The Trouvé Lamp (Fig. 46) consists of an outer
vase within which is a small stationary bell, and in
that is hung a basket of carbide, which may be raised
or lowered by a handle seen on the right of the lamp.
As gas is produced, the water is driven out of the
inner bell and rises in the outer vase. The basket

of carbide must be occasionally lowered a little
deeper into the water.   A layer of oil on the water
within the bell, on the principle of d'Arsonval's

FIG. 46.   Trouvé Lamp.   FIG. 47.   Türr Lamp.

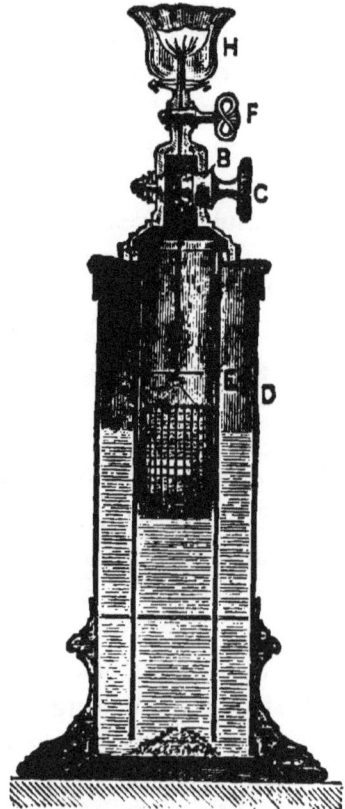

generator, would certainly improve the steadiness of
action of this lamp.

In the **Türr** lamp this has been done (Fig. 47).

The lamp is somewhat larger than the **Trouvé,** and the basket of carbide is suspended by a cord, but otherwise the arrangement is the same.

The lamp devised by Captain **Nou** is in reality a small gasometer and generator (Fig. 48).

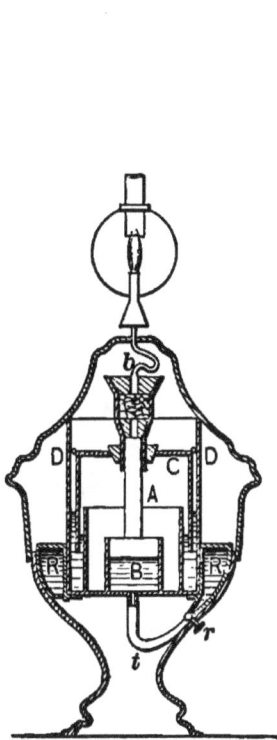

FIG. 48. Nou Lamp. FIG. 49. Claude and Hess Lamp.

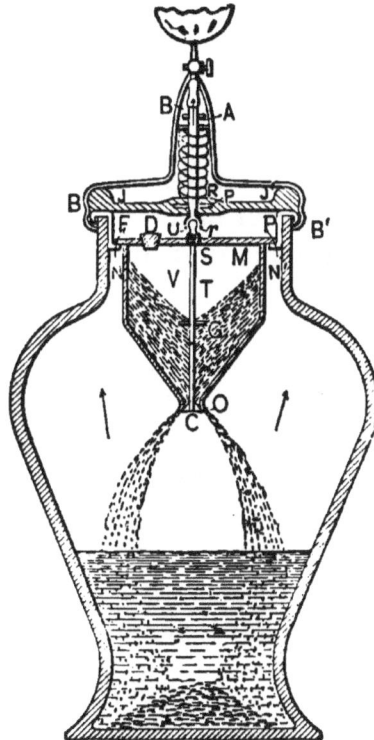

It is said to be very steady in action, but large and unwieldy. Within the outer case is a small gasometer (D) with annular water-seal and the usual bell

(C).   As the bell rises and falls a moulded stick of carbide (A) dips into and is withdrawn from the cup of water (B).   R is a reservoir, from which B draws its supply.

Lamp of Claude and Hess.—This lamp (Fig. 49) has given excellent results, and were it not for the doubtful keeping qualities of pulverized carbide, as well as the uncertainty of getting it of good quality in the first place, would be the best lamp yet devised.

Its construction is simple.   An outer vase contains water in its lower portion, while within its neck is suspended a funnel for holding the pulverized carbide.

A conical plug (C) closes the opening of the funnel. From it a rod passes through the cover of the funnel and is attached to the centre of a thick rubber diaphragm (J, J), which forms the air-tight cover of the lamp.   Above the diaphragm a spring (R) presses down upon it.

When the diaphragm sinks, the plug C is pushed down, allowing carbide to fall into the water, and, as soon as gas is generated, the ensuing pressure pushes up the diaphragm and stops the hole through which the carbide issues.   The gas reaches the burner through the central tube around which the spring is coiled.

A continuous slight working up and down of the diaphragm takes place as the lamp burns, which dusts the carbide slowly and regularly into the water.

The Buffington Acetylene Lamp consists of a generator and burner arranged in the form of a student-lamp. The upper half of the generator contains a water-reservoir, which is provided at its bottom with a needle-valve.

The carbide, enclosed in a cartridge, is introduced into the lower part of the generator by removing the bottom. A water-seal, contained in the removable part, prevents a dangerous rise of pressure.

The lamp is started, stopped and regulated by means of a small lever on top of the generator, which controls the water-supply.

The lamp burns without odor, and the light is very satisfactory.

The Electro, a compact and light acetylene bicycle lamp, is shown by Fig. 50.

The spherical reservoir is filled with water, which, by means of a needle-valve, is allowed to drop upon the cartridge of carbide contained in the lower cylinder.

The gas is consumed in a small burner located in the focus of a parabolic reflector.

The lamp is charged by unscrewing the milled bottom portion of the generator, inserting a cartridge containing about an ounce of carbide, in grains, and filling the reservoir with water through a hole in the top.

It is intended that the gas shall be generated as used, no gasometer or other compensating device being used.

A fairly close regulation may be obtained by means of the needle-valve.

The quantity of carbide is so small that its temperature is not seriously raised by the disengagement of acetylene. The lamp is designed to burn for three hours at each charging.

It seems only fair to state that acetylene lamps are regarded with suspicion by many insurance com-

FIG. 50.

panies, and that some refuse risks on property in which they are operated.

The intending purchaser of any acetylene apparatus should therefore assure his position by obtaining the consent of the company which writes his insurance to the use of the generator or lamp which he expects to install.

THE ordinary house gas-burners, consuming from 2 to 10 feet of gas per hour, will not serve for lighting by means of acetylene. While an ordinary 5-foot burner with a good quality of illuminating gas is rated at 16-candle power, the consumption of a half-foot per hour of acetylene properly burned gives a flame of fully 25 candles.

Acetylene is a gas so rich in carbon, moreover, that special forms of burners are necessary for consuming it without smoke and with a maxim of luminosity.

A half-foot burner of the ordinary type gives a fairly good light with acetylene for a time, but it soon becomes clogged with a deposit of carbon; and, while the flame when burning at a maximum is very satisfactory, it is not so bright as that given by special burners, nor can it be turned down without smoking.

A burner which aims to avoid the deposition of carbon in the gas-outlets is made as shown in the section, Fig 51.

In it the gas-exits are about three-eighths of an inch apart and are inclined towards each other at an angle of 90°. The exits themselves are very small, but a counter-bore enlarges their outer orifices to a diameter of about a thirty-second of an inch.

The issuing jets of gas impinge on each other, producing a flat flame. It is expected that the rush of gas through the air shall carry with it sufficient oxygen for perfect combustion, and that the enlarged outer orifices shall not be clogged by a slight deposit of carbon.

This burner, in spite of these precautions, deposits carbon on that part which is between the jets, and which is comparatively cool.

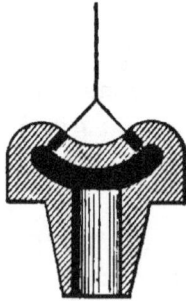   

FIG. 51.   FIG. 52.

The carbon deposit grows in the course of an hour to such size as to deflect the jets, when the flame begins to smoke.

A better form has two slender tubes which, issuing from a common base, curve toward each other (Fig. 52).

The flame is of the same character as that of the last-mentioned burner, but the deposit of carbon is prevented so long as the burner is not turned down low.

In other burners these jets have variously shaped apertures (Fig. 53), but none causes a sufficient mixture of air with the flame to give the best results.

Various expedients have been suggested for dilut-

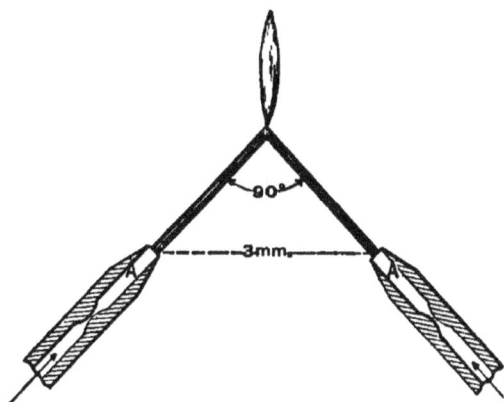

FIG. 53.

ing the gas. Among early experimenters the bold attempt was made to mix air with the acetylene in the gas-holder, but this having resulted in several accidents through the ignition of the explosive mixture so produced, carbonic acid and, afterward, nitrogen were tried.

The last gas gave the best results, but was troublesome to prepare and increased the expense of operation.

Finally, however, by adopting the principle of the Bunsen burner (Fig. 54), it was found possible to mix any desired proportion of air with the gas as it issued from the burner. A further development consisted in making each burner double, with jets inclined toward each other, as in the types last mentioned. This burner, which has been patented in this country, is nearly all that can be desired.

Its flame is intensely bright, it can be turned down without smoking, and deposits no carbon unless left turned down very low. In that case a little ring of hard carbon is likely to form at the exits of the jets, but the burner, even then, does not smoke.

The burners are of the form shown in Fig. 55, being of brass, with so-called lava tips. The tips are veritable little Bunsen burners. A central small aperture leads from the base of the tip to a point opposite the constricted neck of the burner, beyond which point the bore is considerably enlarged. From the constriction, four inclined apertures enter the enlarged bore.

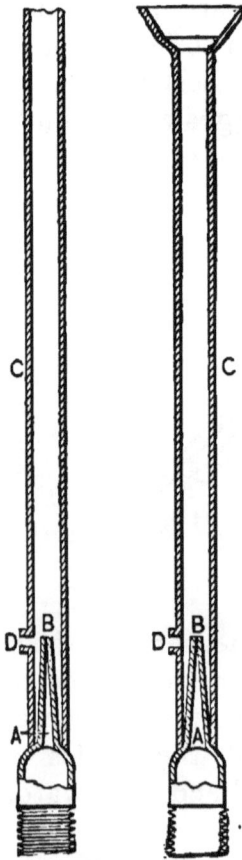

FIG. 54.

From this construction it follows that the gas, in issuing from the small jet into the larger bore, causes a slight reduction of pressure therein, which induces a flow of air through the side openings.

The incoming air and gas mingle and burn on leaving the tip. The two jets, meeting each other, flatten out into a fish-tail shape, and in so doing excite a still further flow of air to their surfaces.

When the burner is turned fully on, the luminous flame does not extend quite to the tips of the jets. The mixture of gas and air next to the jets is not at a suitable temperature for depositing carbon.

The burner illustrated in Fig. 55 is the invention

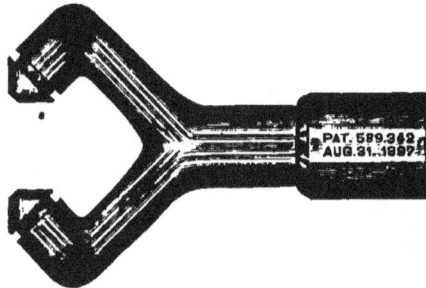

FIG. 55.

of Mr. Edward J. Dolan, of Philadelphia, who has devoted much time to experimental work on the combustion of acetylene. Other forms of burners invented by Mr. Dolan are illustrated in the following figures:

Fig. 56 shows a burner which has a lava tip (A) provided with a deep, vertical slot (B). A small passage (C) delivers the acetylene to the slot (B) at an acute angle. The gas impinges on the wall of the slot (B), and, in its passage to the top of the burner, draws along with it a current of air. The mixture of air and acetylene burns in a fish-tail flame without smoke. The gas-orifice is below the point at which carbon can be deposited.

Fig. 57 shows the same burner having two gas-outlets. It is probably a better form.

Fig. 58 is a burner for consuming a larger quantity of acetylene than is generally burnt in one unit. The tip (G) of this burner has two small Bunsen type of jets inclined toward each other, and, in addi-

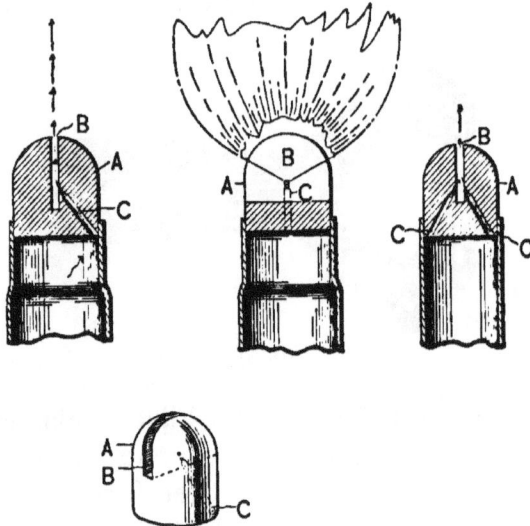

FIG. 56.   FIG. 57.

tion thereto, is provided with a vertical gas-outlet situated between the other two.

A gas-heater by the same inventor is shown in Fig. 59.

The entire structure consists, substantially, of two annular castings, of which 1 represents the base and outer portion, and 2 represents the inner ring, form-

ing the completed burner. The inner ring screws
into the outer by a suitable thread, as at 6. There
is an annular gas-chamber (7), formed by suitable circular slots cut in these two rings, which, when joined

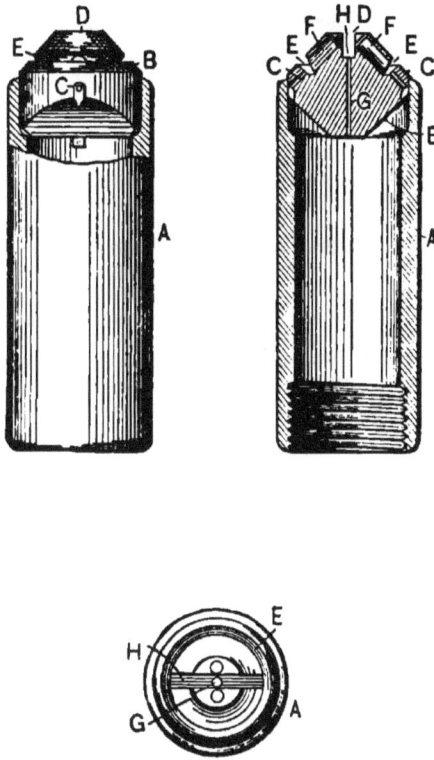

FIG. 58.

together, make the chamber indicated. This communicates with the gas-supply pipe (3). The upper
ring (2) is cast with a series of sectors (8) projecting

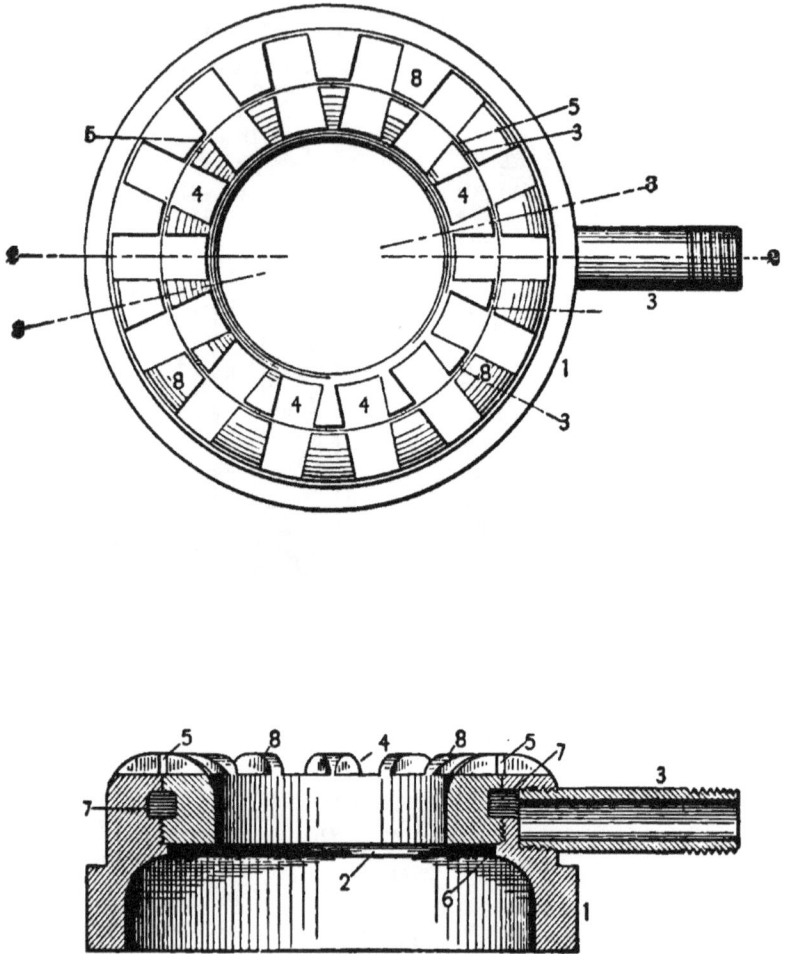

FIG. 59.

above the horizontal surface of the ring. They have between them openings (4) extending down to such horizontal surface, and also transverse slots (5) communicating with the openings or channels (4). The small gas-openings (3) are arranged in the centre of the transverse slots (5), as will be readily seen from the plan, the inner ring (1) being suitably cut for that purpose. Of course these gas-openings can be arranged in either ring, or may be formed between the joined rings. By simply unscrewing the inner ring, the apparatus is readily cleaned.

In operation, air is mingled with the gas in the slots (5), and combustion occurs only above the upper surface of the joined rings, the gas burning there when mingled with a suitable quantity of air to form the desired flame. The material ordinarily employed for the structure is iron, but other materials may also be used.

The ordinary Bunsen burner, made considerably smaller than usual and provided with an adjustable air-inlet, gives a long, slender, luminous flame, which is said to be very effective for groups of lights.

Unless the outer orifice of the Bunsen burner is smaller than a quarter of an inch, the flame snaps back and burns in the tube. With a larger supply of air, the true, colorless Bunsen flame is produced from acetylene, and is much hotter than that of ordinary gas. It may be used with advantage in the laboratory for many operations which usually require the blow-pipe.

Almost every conceivable form of burner and

tip has been tried, but it seems unnecessary to describe or figure those which have been found unsatisfactory.

It may be predicted with a reasonable degree of certainty, from what we already know of the properties and composition of acetylene, that it will be necessary, in all successful burners, to use the Bunsen principle for mingling air with the gas as it leaves the jet.

Many forms of burners have already been designed on this plan, which allows considerable latitude in its application.

A rather wide experience with acetylene burners has shown that the stop-cocks very rapidly cut and become leaky. This is partly due to the deposition of lime carried over from the generator, and partly to the formation of acetylide of copper in the cock itself.

It is advisable, and indeed almost necessary, to occasionally remove the plugs of the cocks and lubricate them with a stiff mixture of bees-wax and tallow, or with some waxy lubricant containing graphite. The graphite composition used upon the chains of bicycles may with advantage take the place of other lubricant.

Cocks which are allowed to become dry, and which turn with difficulty in their sockets, soon after begin to leak and emit the peculiar odor of acetylene; and after this happens there is no cure but regrinding.

AUTHOR'S EXPERIMENTS

THE author's experiments with generators began with a small apparatus constructed on the principle of the hydrogen lighter (Fig. 29), in which a reservoir was partly filled with water having a layer of oil about six inches deep floating thereon. A bell descended into the outer vessel. It had an opening through which a perforated can, filled with calcium carbide, could be introduced and adjusted in height by means of a rod passing through a stuffing-box. This affair worked remarkably well on a small scale. The oil, into which the carbide was lifted by the rising bell, effectually displaced the water and stopped the evolution of gas. When one or two burners were in operation, the bell descended until the carbide was partly immersed in the water, the degree of immersion being such as would supply the gas at the rate used.

In this respect the device was very satisfactory, and was quite self-regulating. The carbide being introduced in a finely perforated can, the amount of lime which escaped into the water was small. The disadvantages were such as to preclude the use of the apparatus for anything more than experimental work.

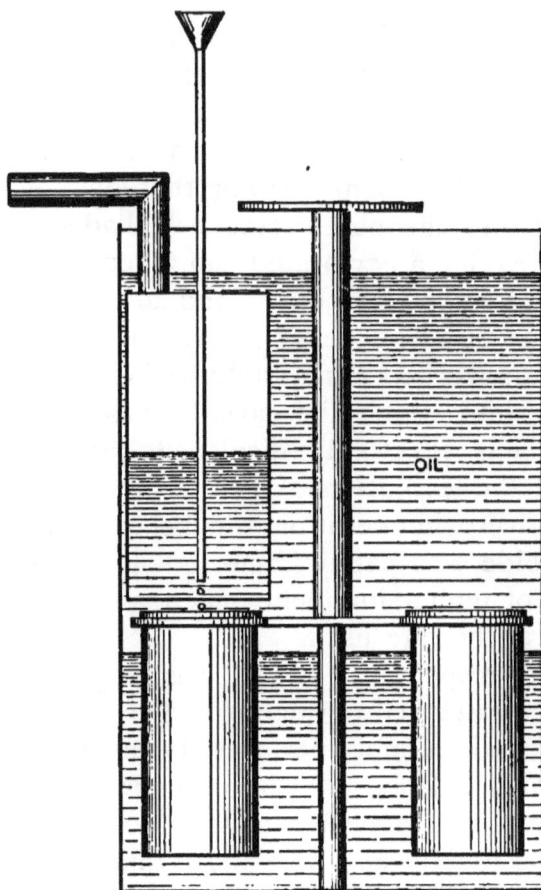

FIG. 60.

In the first place, the introduction of fresh carbide involved the loss of the remnant of gas contained in the bell.   Secondly, the petroleum-oil used absorbed gas readily, and as easily gave it off, causing a strong odor of acetylene to issue from the machine.   This could be overcome, however, by having the oil within the bell only and never permitting the edge of the bell to rise to a height sufficient to allow of its escape.

Thirdly, the use of strong brine was necessary in winter to prevent the freezing-up of the machine, which was kept out of doors.

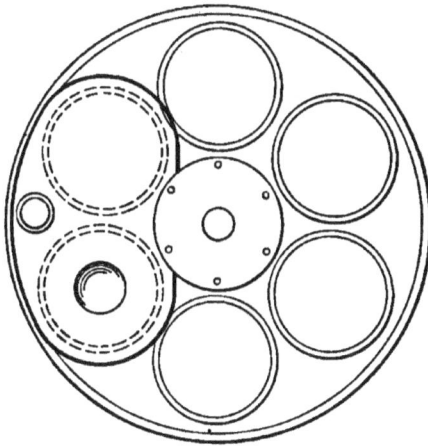

FIG. 61.

The use of oil in this way must have occurred to many, and the discovery, only recently, that the identical machine had been devised a year earlier in France by d'Arsonval was not surprising.

The next expedient tried was the immersion of the carbide in oil at the bottom of a reservoir, and the addition of the water thereto in a small stream, which, sinking through the oil, attacked the carbide. The carbide was contained in a series of cans, any one of which could be brought under a stationary bell in turn, and there receive the water which entered the bell through a small pipe, and which had its flow regulated by the rise and fall of a gasometer connected to the generator (Fig. 60).

Uneven production of gas, bad odor, and general mussiness caused this plan to fail in practical application.

The plan is unusually attractive, and, with improvements, may yet offer a simple solution of the problem of the generation of acetylene.

Experiments showed that it was not only desirable, but necessary, to have any machine for generating acetylene in sufficient quantity for domestic lighting removed and quite isolated from inhabited buildings. Aside from the possible element of danger, there was the disagreeable odor, which is quite inseparable from the use of calcium carbide. Very little smell is noticed in charging a properly constructed machine, but the removal of the spent lime, always charged with a strong odor of gas, is, to say the least, unpleasant unless conducted in the open air. The gaseous odor has a peculiarly lingering and diffusive character. A little of it goes a long way, and to some it is intensely disagreeable. The isolation of the generator means that it must be placed in a separate building sufficiently protected from the weather, or so heated, as to prevent the freezing-up of the machine, or it must be placed in a vault underground.

The last-mentioned proceeding seems the more

desirable, since all danger of freezing is removed when the surface of the water in the generator or gas-holder is three feet or so under the surface of the ground; and in addition to this advantage is that of avoiding the erection of a separate and probably unsightly or undesirable building.

The only disadvantage arising from this disposition of the generator is the difficulty of removing the lime resulting from the action of the water on the calcium carbide.

The so-called " chute " type of machine lends itself particularly to underground use.

The first machine of this kind experimented with consisted of a sheet-iron cylinder three feet in diameter by seven in height. A chute entered the lower portion of the cylinder, which was buried in the ground.

From the bottom of the cylinder a two-inch pipe led with a gentle curve to the surface of the ground, where it was surmounted by a common pump. An inverted bell within the cylinder formed the gas-holder.

It was found necessary to add a deflector within the cylinder, since the evolution of gas was sometimes so rapid that bubbles would be driven to the sides of the cylinder and rise outside the bell.

The carbide was at first dropped down the charging-pipe by hand. As the capacity of the bell was twenty cubic feet, and as it was undesirable to allow it to become entirely empty of gas before refilling, it was customary to add the carbide in charges of about three pounds, which were simply dropped from a tin-can into the charging-pipe.

A lively evolution of gas immediately followed, lasting about ten minutes, when the carbide was all reduced and the bell full of gas.

After using a hundred pounds of carbide, the precipitated lime was pumped out into a couple of galvanized-iron cans, from which, after the lime had settled to the bottom, the water was decanted back into the generator.

An agitator was incorporated into the machine. The central part of its lower extremity was fitted to the bottom of the pump suction-pipe, thereby preventing lime from settling and becoming impacted therein. The handle of the agitator passed up through the bell, for which it formed a guide. In use, the agitator was gently revolved a few times and then lifted up a few inches. The pump was immediately put to work, and no difficulty was found in removing the lime in the form of a thin paste.

Later developments showed the desirability of making the operation of the machine continuous. In order to accomplish this end, an automatic device was added which emptied the contents of a sealed can of carbide into the charging-tube whenever the bell sank to a certain point. Figs. 62, 63.

At first the expedient of soaking the carbide in kerosene was tried, and the charging-pipe, or chute, was, after the manner devised by Hospitallier, partly filled with kerosene, which floated upon the water.

Two objects were attained by this procedure.

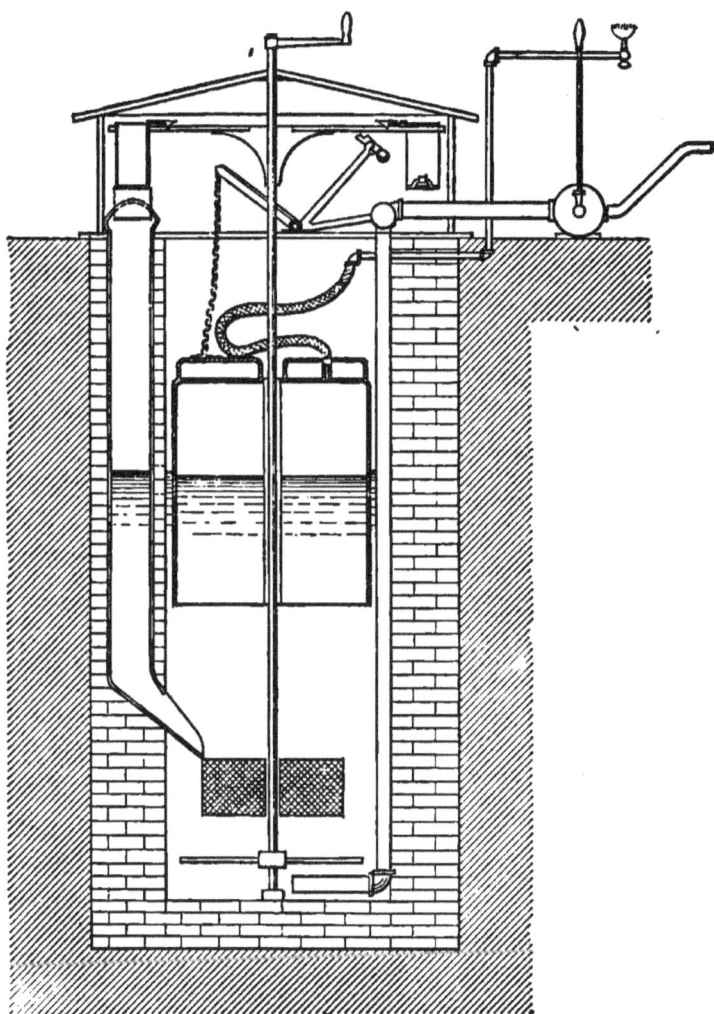

FIG. 62.

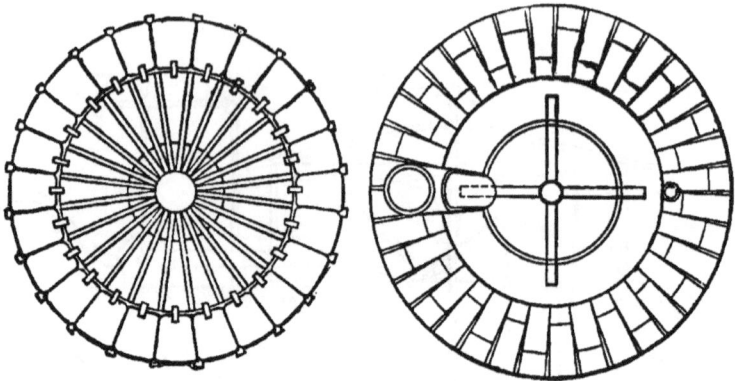

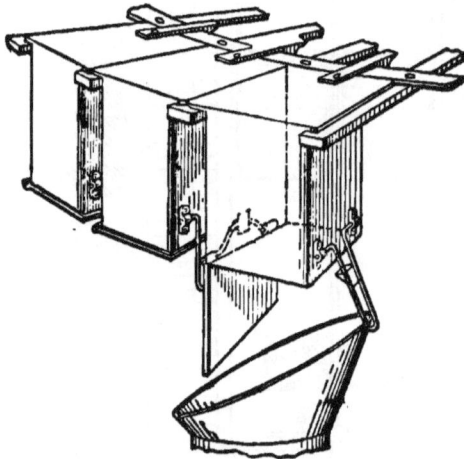

FIG. 63.

The carbide in the original package was protected by the oil from the action of the atmosphere, so that, when a fresh can was opened, the pouring in of a gallon of oil rendered the use of an air-tight cover unnecessary.

The layer of oil in the charging-pipe, by forming a coating on the carbide in its passage down the chute, delayed the action of the water until the generator was entered.

Continued experiment showed that the oil which entered the generator with the carbide was so changed in structure that it formed a thick, unsightly scum on the surface of the water under the bell, which retarded the disengagement of gas and was a most disagreeable feature when it became necessary to clean the machine. This scum finally became so troublesome that the use of oil was reluctantly abandoned.

The cans containing the carbide on the distributor were furnished with rubber washers, on which the cover bore, making an air-tight joint after the fashion of the ordinary glass fruit-jars. It was also found that the descent of the carbide through the chute was so rapid that practically no gas was given off until the generator was reached.

The distributor above mentioned consisted of a horizontal disc turning freely on a rod which passed through its centre. It was pierced near its edge with a row of holes, each of which held, by means of a bayonet-joint, a carbide can four inches in diameter by seven inches deep. The cans were held top down, and the cover of each was held shut against a rubber washer by means of a spring latch.

A ratchet and pawl, operated by the fall of the gasometer bell, brought each can in turn over the charging-pipe, where its latch was tripped, and the contents dropped down the chute into the generator.

A sufficient number of cans was provided to hold the contents of a one-hundred-pound package of carbide, which, on being opened, was at once distributed among them and sealed by closing the air-tight covers.

Once a week the empty cans were removed from the distributor and replaced by full ones. A glance showed which were empty, since the covers hang down after the latch is tripped.

At the time that this machine was constructed, it was thought to possess some advantages over others, inasmuch as it was non-freezing, had no valves nor cocks, could not accumulate pressure, and was automatic in action.

It was seen, however, after due consideration, that in ignorant hands there was one possibility of accident. If, in pumping out the lime, the level of the water in the gasometer was unduly lowered, it might happen that the bell would reach the bottom of its travel. Further pumping would create a vacuum, tending to draw the flame of any lighted burners back into the piping and eventually to the bell. That this can happen seems doubtful, but it has been said to have occurred under certain conditions. At all events, the lights would go out and air would enter the system. Continued pumping would lower the level of the water below the edge of the bell, allowing air to enter. If, under these circumstances, the generator was refilled with water without having taken the precaution to draw off the mixture of air and gas from the bell, the conditions would favor an explosion if a flame was brought into proximity with the smallest leak or imperfect joint.

In order to guard against just such conditions, various forms of machine were devised.

Some had the generator and gasometer separate, while others provided for the removal of the lime by lifting-out cans, into which the precipitate was directed.

After many experiments, the original form of machine as last described was finally adopted, and the very simple expedient tried of making a small hole through the wall of the pump suction-pipe at the lowest intended water-level, since it was found that the pump removed the lime perfectly and with less labor than any other method.

The generator finally took the form shown in the last engraving, after passing through the intermediate stage shown in Figs. 62 and 63.

The gas generating and holding portion of the acetylene machine is contained in a cylindrical cistern of iron or brick, of an area determined by the amount of gas required.

The cistern is partly filled with water, the surface of which is below the line of frost.

A gasometer-bell occupies the upper portion of the cistern, which it nearly fills, and rises or sinks in the water as gas enters or is drawn from it.

This bell is provided at its upper part with an air-tight, annular flotation-chamber, the use of which

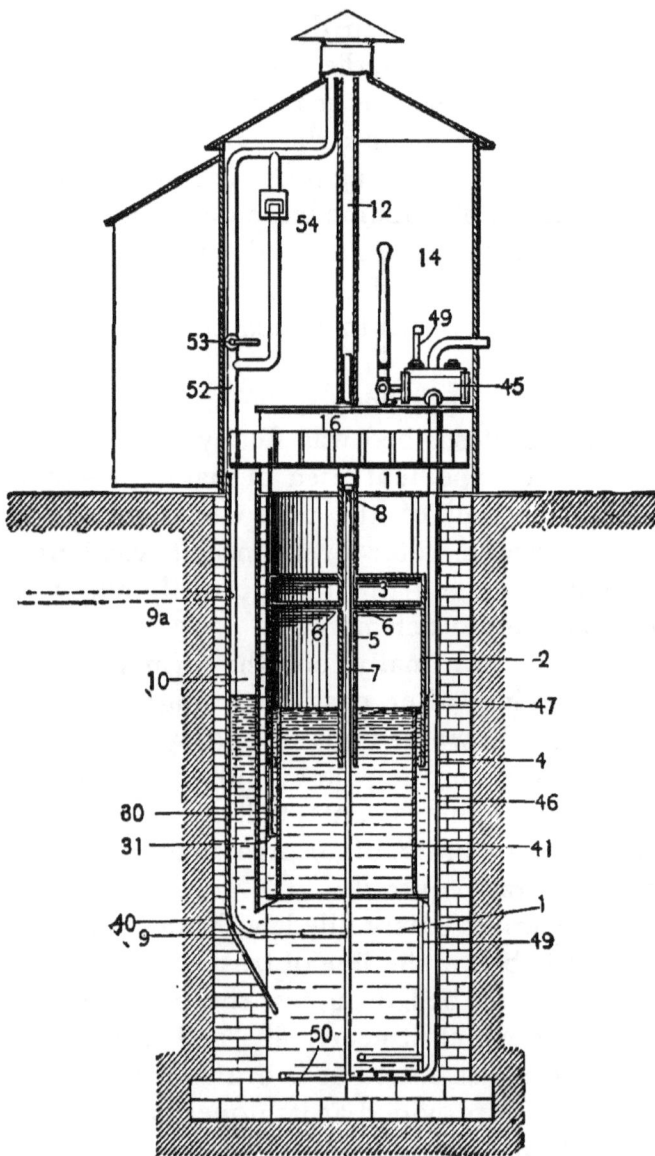

FIG. 64.

will be described later, and has at its lower edge a
heavy rim of iron, for the purpose of giving sufficient pressure to the gas contained therein.

The bell contains a central tube, open at the bottom, which projects for some distance above the top
of the flotation-chamber and serves as a duct for
leading the gas from the bell to a second smaller
concentric tube fixed in the cistern.

This second tube has openings above the water-
line of the cistern, and is connected with the system
of piping and burners where the gas is used. Its
upper end telescopes into the tube of the gas-holder,
for which it forms a guide.

A "chute," contained in the wall of the cistern,
directs the calcium carbide dropped into it to the
generating portion of the machine, located at the
bottom of the cistern.

The cistern is covered by a drum of iron, into
which the gasometer may rise. This drum is closed
on top, but is provided with a vent-pipe leading into
the ventilator of the small iron building which covers the whole machine.

On a circular track around the circumference of
the drum travels a series of small cans.

Each can has a hinged bottom provided with a
rubber washer and held shut by a latch. Since the
cans, when closed, are air-tight, the carbide within
them is entirely protected from the air until the moment when it is needed, at which time the gasometer
bell, in its descent, by means of a ratchet-and-pawl
mechanism, carries a can of carbide over the chute,

trips the latch and drops the contents into the bottom of the cistern.

The gas evolved by the reaction between the water and the calcium carbide rises through the large quantity of water above it, being directed into the bell by means of the deflector. It reaches the bell cool and thoroughly washed.

As the bell rises and falls, carbide is automatically fed to the generator, canful by canful, so long as any remains.

The removal of the deposit of lime, which is necessary only at long intervals, is effected by means of a pump and a suction-pipe which reaches the bottom of the cistern.

A small hole is bored through the suction-pipe at the normal level of the water in the cistern.

The process of removing the lime consists in first adding water to that in the cistern until the level is raised a couple of feet.

The pump is then operated as long as it removes lime and water.

When the water in the cistern has been reduced by pumping to its normal level, air enters the small hole in the suction-pipe, stopping the action of the pump.

Consequently, the water-line can never by carelessness be brought below the point intended.

The operation of the ratchet, and consequent feeding of carbide, when too much water is contained in the cistern, would be undesirable, causing over-production and waste of gas.

This is prevented by the flotation-chamber contained in the bell, for, since the bell cannot sink below the water-level, it cannot fall far enough to operate the can-moving mechanism unless the water-line is normal.

The gas-delivery pipe has a branch extending into the ventilator of the covering-house, with a cock for allowing the gas contained in the bell to be drawn off, if desired.

A loop of pipe around the cock contains a mercurial seal, which automatically discharges any surplus gas in case its pressure exceeds four inches of water.

The covering-house protects the machine from the weather, and prevents access to the working parts by unauthorized persons. It also provides ample storage-room for a supply of calcium carbide, and a place in which the carbide may be transferred from the original packages to the air-tight cans in which it is supplied to the machine.

A sufficient number of cans is provided, in addition to those in use, to hold the contents of a one-hundred-pound package of carbide.

The carbide cans are held in place on the charging-wheel of the machine by a simple fastening, which permits the empty ones to be easily removed and full ones substituted.

It is necessary to pump out the lime only twice a year when the machine is supplying the ordinary demands for a dwelling-house in which there are from twenty-five to thirty burners.

In suburban places, it answers perfectly to dispose of the lime by pumping the contents of the machine into a hole dug in the ground. The water disappears in a short time by seepage leaving the lime at the bottom of the hole, which is then filled with the earth previously taken out. A hole three feet deep is ample in most soils. In many cases, a smaller one will do as well.

Instead of making the outer case of iron, it may be built up of brick laid in portland cement, or it may be formed of earthern pipes, which may be obtained of any diameter up to three feet.

In giving this machine, which contains about forty cubic feet of water, its initial charge of calcium carbide, it is a noteworthy fact that the first four or five pounds of the substance thrown into the chute fail to cause more than two or three inches rise of the gasometer bell.

In other words, the solubility of the gas in the water is so readily effected that, until a considerable degree of saturation has taken place, very few of the bubbles of gas which are generated at the base of the machine reach the surface of the water.

The action reminds one of the "singing" stage of a kettle which is set to boil, where the bubbles of steam generated at the bottom of the kettle are condensed by the cooler upper portion of the liquid before they reach its surface.

Two inferences may be drawn from these facts: First, that the renewal of the water should be avoided as much as possible, or that a saline solution which absorbs only five per cent. of its volume of gas should be used.

Second, that a source of danger from explosions exists in this body of gas-charged water.

Suppose a leak had been discovered in the bell of the gasometer. A careless individual might let all the gas escape, and then, in order to easily reach the bell, might draw it out of the water, letting air enter as it was raised. He might then, without removing it from the liquid, attempt to mend the leak with a soldering-copper or a blow-pipe, under the impression that there was no gas in the bell. In reality, as the bell was raised, the pressure on the water would be diminished, so that a certain amount of gas would be set free and would mingle with the air in the bell, forming an explosive mixture.

On the other hand, in the usual routine of gas making and using, the large body of water is a decided safeguard, preventing, as it does, any appreciable rise of temperature of the carbide or gas, and therefore guarding against any decomposition, while at the same time it gives the gas a thorough washing on its way to the bell.

The gas generated from the lumps of carbide leaves them in a continuous stream of small bubbles. There is no violent, sudden or explosive action visible. Although the evolution of acetylene is rapid, the gas is broken up into very small bubble units. From this it follows that, while the surface of the water, as seen when the bell is removed, is thrown into violent commotion, and is in every part in brisk ebullition, the water itself is not thrown into the air, but rolls over and over as the bubbles leave its surface.

CONCLUSION

AFTER a careful study of the United States patent specifications of the acetylene generators, which will be found at the end of this volume, the author is impressed with the curious fact that they may all be included in the first two classes.

The greater number are of that variety in which the carbide is contained in a closed receptacle, to which water is fed in small quantities.

The remainder are of the type which the French call "briquet hydrogen." They have the carbide suspended in a basket within the gas-holder, with means for alternately immersing and withdrawing it from the water contained therein. With the exception of one machine, which occupies an intermediate position, there is none in which the carbide in small charges is dropped into a large quantity of water. This class of machine, which has been named the "chute generator," offers certain advantages which cannot be obtained with the others.

The French and German experimenters upon acetylene, who have earnestly sought the best means of gas-production, report uniformly in favor of the chute generator. When properly made and installed, however, this machine is rather expensive.

147

This is especially so when means are provided for the periodical removal of the deposited lime, and when the generator is made of sufficient capacity to supply gas continuously over a prolonged space of time.

The American public, as a whole, is unusually well-informed concerning the development and progress of recent inventions. It wishes, moreover, to encourage the introduction of such novelties as contribute to its comfort, save its time, lessen its labor, improve its health, or in any way render more easy the pursuit of happiness.

That the American is by nature an inventor is a proof that he meets with support from his fellows. Coupled, however, with the desire to look with favor upon all that which is new and meritorious, there is constantly in the demand of the public a cry for cheapness.

The advent of the acetylene light has satisfied this requirement with an illuminant of unprecedented excellence. That having been accomplished, there seems to be no good reason why further demands for a cheap generator, unless entirely consistent with safety and durability, should meet with encouragement. One of the most attractive features of acetylene production is the ease with which the gas may be generated in quantity sufficient for a demonstration of its properties. It seems unfortunate, however, that the performance of the crudest and most flimsy generator should be, to the uninitiated, so satisfactory, for while, as has been said, the public is in sympathy with new devices, it lacks the technical knowledge to differentiate between the good and poor machine upon their respective showings.

The inventor is little to blame who, stimulated by competition and the inability to sell any but the cheapest generator, reduces the size and cost of his machine to the lowest possible limits.

Acetylene is, until "something happens," a substance so easily managed, so capable of control and apparently so unlikely to manifest its latent affinities, that the vigilance of the experimenter becomes relaxed. He attempts things in his increasing confidence which he would consider dangerous had his experience been more varied.

The very simplicity of acetylene production, which renders the construction of cheap and unreliable generators possible, has acted as a check to the further development of the art.

The public, in making cheapness the most important feature of a machine, is, perhaps, also blameless, in default of more general technical information.

As acetylene becomes more generally used and the public becomes better informed concerning the properties of the gas and of the carbide, it will be placed in a position to decide for itself how much risk it is warranted in taking.

In the meantime, the responsibility should be made to rest with the fire insurance companies. It does, in fact, rest there now, for it is true that safe and efficient generators are not lacking. It is also true that the insurance companies, through the boards of fire underwriters, employ experts to pass judgment upon the machines submitted to them.

Unfortunately, the experts in some cases have not paid that heed to the increasing use of acetylene generators which its importance would seem to demand. Any laxity in this matter will, however, be corrected when the companies who write insurance upon property containing machines of doubtful safety learn by costly experience the error of their ways.

As a guide to the experimenter who is seeking to devise an acetylene generator, the requirements of the New York Board of Fire Underwriters are here reprinted.

These requirements, in the main well devised to prevent the use of dangerous machines, are certainly very exacting, but in the present state of the art a conservative attitude is the only one which is quite safe.

As more experience is gained, the requirements will, no doubt, be changed from time to time, in order to keep abreast of the developments.

As a temporary measure, at least, they seem all that can be desired.

REQUIREMENTS

OF

THE NEW YORK BOARD OF FIRE UNDERWRITERS

FOR THE

Installation of Acetylene Gas Generators

AND FOR THE

STORAGE OF A LIMITED SUPPLY OF CALCIUM CARBIDE.

---

1. Plans and specifications in detail of Acetylene Gas Generators must be submitted to this Board for approval, and a copy of the same placed on file in this office. If the plans are approved, a special examination of the generating apparatus will be made (at the expense of the applicant), and if it is found to be in compliance with the following requirements, a certificate of approval will be granted :   ,

2. The generating apparatus must be located in an outside, fireproof and well-ventilated building, where it will not be an exposure to any adjoining property. The buildings in which generators having a capacity of more than twenty-five pounds and not exceeding one hundred pounds of calcium carbide are placed, shall not be located within ten feet of any other building ; and the buildings in which generators having a larger capacity than one hundred pounds and not exceeding five hundred pounds of calcium carbide are placed, shall be restricted to a distance of not less than twenty-five feet from any other building, and these shall have the constant supervision of a competent person.

3. The dimensions of the generator building must be confined to the requirements of the apparatus, and the limited

151

supply (hereinafter mentioned) of surplus calcium carbide, which must be packed in water-tight metal cans, and said buildings shall be located as follows :

For generators with capacity of more than twenty-five pounds and not exceeding one hundred pounds of calcium carbide, and, in addition, one hundred pounds of surplus carbide —not less than ten feet from other buildings.

For generators with capacity of over one hundred pounds, and not exceeding five hundred pounds of calcium carbide, and, in addition, not over five hundred pounds of surplus carbide—not less than twenty-five feet from other buildings.

The storage of calcium carbide on premises, other than in generator building, is absolutely prohibited.

4. In constructing the building, dryness and ventilation must be secured. To meet these requirements, the floor must be raised above the grade on which the building is located, and suitable drainage provided. Ventilation is to be obtained by air passing from the outside of building through holes at the floor and through a pipe at least six inches in diameter, at the roof. The said pipe must extend at least four feet above the roof, and must be topped with a guard-cap, and if there be any building within ten feet of said pipe, then the ventilating pipe must be carried four feet above the roof of the higher building.

5. The maximum pressure of gas stored in a gas-holder shall be limited to eight inches of water, and both the generator and gas-holder shall have water safety-seals (not to exceed the same limit) in connection with escape pipes of not less than one and one-half inches in diameter. The escape pipes must be connected above the roof with the ventilating pipe of the building in which the generator is located.

6. A generator in which the gas is both generated and stored (the maximum pressure of which shall not exceed five pounds per square inch) and having no water-seal, shall be tested to withstand a hydraulic pressure of twenty pounds per square inch. The generator shall have a pressure-gauge, also

a safety-valve (not less than one and one-half inches in diameter) adjusted to release the pressure of gas should it rise above the prescribed maximum limit. An escape pipe must be affixed to the safety-valve, and connected above the roof to the ventilating pipe of the building in which the generator is located. A certificate attesting the hydraulic test is to be placed on file in this office.

7. The pressure of the gas shall be regulated at the generator or gas-holder, so that it will not exceed four inches of water on the pipes inside of the building to be lighted. A mercurial seal, set to that pressure, must be attached to the supply-pipe at the generating building. A stop-valve shall be placed on the supply-pipe at the place where it enters the inside of the building to be lighted.

8. All generators and gas-holders shall be connected by at least one and one-half inch escape pipes and stop valves with the ventilating pipe at the roof of the building, through which the gas can be conveyed and discharged with safety on the outside of the building.

9. Generators must be constructed so that they can be charged with calcium carbide at all times without allowing the gas to escape into the building.

10. Generators shall be filled with calcium carbide by daylight only, and all generating apparatus must be in charge of persons who are familiar with their operation and are fully competent to manage them under all circumstances.

11. No artificial light, except a wire-guarded incandescent electric light, may be used inside of the building in which the gas is generated, and no heat except low-pressure steam.

12. All acetylene gas generators, and all receptacles containing acetylene gas, shall be made of iron or steel throughout.

13. The residuum of the calcium carbide when removed from the generator must be deposited in a safe location outside of the building apart from any combustible material.

LIQUID ACETYLENE.

14. The storage of liquid acetylene in any building, or the use of liquid acetylene gas, is absolutely prohibited.

Calcium Carbide and Acetylene Apparatus

CARBIDE AND ELECTRIC FURNACES.

Number.	Date.	Inventor.	Title.
492,767 Reissue 11,473	Jan. 28,1893 Feb. 26,1895	Acheson, Edward G..	Production of artificial, crystalline, carbonaceous materials.
541,137	June 18, 1895	Willson, Thomas L.....	Calcium carbide process.
541,138	June 18, 1895	Willson, Thomas L.....	Product existing in crystalline calcium carbide.
551,461	Dec. 17, 1895	Clarke, William C......	Art of producing calcium carbide.
Reissue 11,511	Oct. 22, 1895	Willson, Thomas L.....	Calcium carbide process.
552,036	Dec. 24, 1895	Böhm, Ludwig K.......	Material for incandescent conductors.
552,890	Jan. 14, 1896	Clarke, William C......	Manufacture of calcium carbide.
555,796	Mar. 3, 1896	Whitehead, C..........	Compound of magnesium and calcium carbide.
557,057	Mar. 24, 1896	Dickerson, Edward N..	Process and apparatus for producing metallic compounds by electricity.
560,291	May 19, 1896	Acheson, Edward G....	Electrical furnace.
562,402	June 23, 1896	King, William R., and Wyatt, Francis.	Process of forming calcium carbide.
563,527	July 7, 1896	Willson, Thomas L.....	Process of producing calcium compounds.
563,528	July 7, 1896	Willson, Thomas L	Process of manufacturing hydro-carbon gas.
571,084	Nov. 10, 1896	Eldridge, Hilliary, Clark, Daniel J., and Wambaugh, Mahlon W.	Composition of matter for manufacturing calcium carbide.
572,636	Dec. 8, 1896	Hewes, James E........	Electric furnace.
578,685	Mar. 9, 1897	Whitney, Edwin R.....	Process and apparatus for producing calcium carbide.
583,408	June 1, 1897	Morehead, James T.....	Manufacture of carbide of calcium.
587,138	July 27, 1897	Roberts, Isaiah L......	Process of and apparatus for manufacturing metallic carbides.

CARBIDE AND ELECTRIC FURNACES—*Continued.*

NUMBER.	DATE.	INVENTOR.	TITLE.
587,343	Aug. 3, 1897	Strong, George S.......	Electric furnace.
587,509	Aug. 3, 1897	Roberts, Isaiah L	Process of and apparatus for making metallic carbides.
588,012	Aug. 10, 1897	Roberts, Isaiah L......	Process of and apparatus for making metallic carbides.
588,866	Aug. 24, 1897	Kenevel, Jeannott W...	Means for manufacturing carbides.
589,592	Sept. 7, 1897	Blum, Sylvain.........	Composition of matter for manufacturing calcium carbide.
589,967	Sept. 14, 1897	Heath, Robert F. S.....	Composition for manufacturing calcium carbides.
590,514	Sept. 21, 1897	Cowles, Alfred H......	Process of producing metallic carbides.
597,945	Jan. 25, 1898	Bradley, C. S	Electric furnace.

GENERATORS.

535,944	Mar. 19, 1895	Dickerson, Edward N..	Process of and apparatus for producing and liquefying acetylene gas.
541,429	June 18, 1895	Dickerson, Edward N..	Process and apparatus for producing gas.
541,428	June 18, 1895	Dickerson, Edward N..	Automatic gas-holder relief-valve.
541,427	June 18, 1895	Dickerson, Edward N..	Apparatus for production of gas.
541,526	June 25, 1895	Dickerson, Edward N..	Process of and apparatus for manufacture of gas.
542,320	July 9, 1895	Willson, Thomas R.....	Process of and apparatus for manufacture of gas.
550,162	Nov. 19, 1895	Dickerson, Edward N..	Gas-governor.
552,027	Dec. 24, 1895	Willson, Thomas L.....	Process of generating gas.
552,028	Dec. 24, 1895	Willson, Thomas L.....	Apparatus for generating gas.
551,815	Dec. 24, 1895	Farnsworth, Ezra	Apparatus for manufacture of gas.
552,048	Dec. 24, 1895	Dickerson, Edward N...	Process of and apparatus for mingling gases.
552,375	Dec. 31, 1895	Jones, Charles C.......	Acetylene gas generator.
552,099	Dec. 31, 1895	Clarke, William C......	Apparatus for generating gas.
552,101	Dec. 31, 1895	Clarke, William C......	Apparatus for generating and supplying gas.
552,100	Dec. 31, 1895	Clarke, William C......	Apparatus for generating gas.
553,443	Jan. 21, 1896	Willson, Thomas R.....	Process of carburetting water-gas.

GENERATORS—*Continued.*

NUMBER.	DATE.	INVENTOR.	TITLE.
553,550	Jan. 28, 1896	Willson, Thomas L.	Process of producing illuminating gas.
553,781	Jan. 28, 1896	Dickerson, Edward N ..	Apparatus for producing gas.
555,149	Feb. 25, 1896	Dickerson, Edward N ..	Process for and apparatus for burning liquefied gas.
555,198	Feb. 25, 1896	Willson, Thomas L.	Process of making and consuming gas.
555,212	Feb. 25, 1896	Dickerson, Edward N ..	Process of and apparatus for producing illuminating gas.
556,115	Mar. 10, 1896	Turney, E. T..........	Process of and apparatus for producing illuminating gas.
556,736	Mar. 24, 1896	Clarke, William C......	Method of and apparatus for generating acetylene.
556,737	Mar. 24, 1896	Clarke, William C......	Method of generating illuminating gas.
556,910	Mar. 24, 1896	Wilkinson, A. W.......	Process of manufacturing gas.
556,911	Mar. 24, 1896	Wilkinson, A. W.......	Process of manufacturing gas.
558,746	April 21, 1896	De Sieghardt, O. T.....	Apparatus for generating and storing acetylene gas.
559,846	May 12, 1896	Gray, G. J., and Hitchcock, W. F.	Apparatus for manufacturing acetylene gas.
560,405	May 19, 1896	Fuller, H. F...........	Acetylene gas generator.
560,549	May 19, 1896	Seward, O. G., Mille, O. E., and Ham, M. J...	Automatic gas generator.
560,784	May 26, 1896	Dickerson, Edward N ..	Valve-locking mechanism for gas generators.
561,208	June 2, 1896	Dickerson, Edward N ..	Automatic acetylene gas apparatus.
561,701	June 9, 1896	Dickerson, Edward N ..	Process of producing acetylene gas.
562,040	June 16, 1896	Sergeant, H. C........	Gas generator.
562,401	June 23, 1896	King, W. R., and Wyatt, F.	Apparatus for generating acetylene gas.
562,911	June 30, 1896	Porter, J. C..........	Acetylene gas generator.
563,457	July 7, 1896	Dickerson, Edward N ..	Acetylene gas generator.
563,980	July 14, 1896	Morley, J. H..........	Acetylene gas generator.
563,981	July 14, 1896	Morley, J. H..........	Acetylene gas generator.
564,684	July 28, 1896	Dickerson, Edward N ..	Gas-mixing device.
565,157	Aug. 4, 1896	Dickerson, Edward N ..	Gas-mixer.
566,660	Aug. 25, 1896	Clarke, H. B..........	Acetylene generator and bicycle-lamp.
566,901	Sept. 1, 1896	Fuller, H. F.....	Acetylene gas generator.
567,641	Sept. 15, 1896	Eldridge, H............	Gas-generating apparatus

GENERATORS—*Continued.*

Number.	Date.	Inventor.	Title.
567,773	Sept. 15, 1896	Rossback-Rousset, F ...	Gas-generating lamp.
569,273	Oct. 13, 1896	Bucher, A. S............	Acetylene gas generator.
569,208	Oct. 20, 1896	Exley, J. H.............	Apparatus for manufacturing acetylene.
571,269	Nov. 10, 1896	Janeway, J. L..........	Process of manufacturing gas.
571,576	Nov. 17, 1896	Porter, J. C............	Gas generator.
572,113	Dec. 1, 1896	Hill, W. P., and H. D..	Acetylene gas generator.
573,996	Dec. 29, 1896	Owen, R. L.............	Gas-distributing apparatus.
573,938	Dec. 29, 1896	Waite, John H.........	Acetylene gas generator.
574,601	Jan. 5, 1897	Casgrain, H. E	Acetylene gas generating lamp.
575,281	Jan. 12, 1897	Buffington, L. S........	Apparatus for generating acetylene gas.
575,474	Jan. 19, 1897	Fuller, Henry F........	Gas generator for acetylene.
575,677	Jan, 19, 1897	Fuller, Henry F........	Method and apparatus for generating acetylene gas.
575,885	Jan. 26, 1897	Fourchotte, M. C. A....	Apparatus for producing acetylene gas.
575,884	Jan. 26, 1897	Fourchotte, M. C. A....	Apparatus for producing acetylene gas.
576,386	Feb. 2, 1897	Voisard, E. P..........	Acetylene gas generator.
576,585	Feb. 9, 1897	Kidder, Moses W.......	Apparatus for generating acetylene gas.
576,826	Feb. 9, 1897	Sergeant, H. C.........	Generator for making acetylene gas.
576,529	Feb. 9, 1897	Addicks, W. R.........	Apparatus for manufacturing gas.
576,827	Feb. 9, 1897	Sergeant, H. C.........	Acetylene gas holder.
576,893	Feb. 9, 1897	Reynolds, D. J.........	Acetylene gas generator.
576,955	Feb. 9, 1897	Deuther, J A..........	Method of and apparatus for generating gas.
577,051	Feb. 16, 1897	Matthews, Charles, Jr..	Acetylene gas generator.
577,803	Feb. 23, 1897	Willson, Thomas L.....	Process of producing and consuming hydro-carbon gas.
577,706	Feb. 23, 1897	Archer, G. S., and Burrington, C. F........	Acetylene gas generator.
577,762	Feb. 23, 1897	Lawrence, R. S	Apparatus for increasing candle-power of gas.
578,055	Mar. 2, 1897	Fuller, Henry F	Acetylene gas generator.
578,847	Mar. 16, 1897	Wilcox, Clementina H..	Acetylene gas generator.
578,972	Mar. 16, 1897	Couper, J. H...........	Acetylene gas generator.
579,702	Mar. 30, 1897	Dickerson, Edward N...	Acetylene gas producing apparatus.
579,689	Mar. 30, 1897	Vincent, J. A..........	Acetylene gas producing apparatus.

GENERATORS—*Continued.*

NUMBER.	DATE.	INVENTOR.	TITLE.
580,624	April 13, 1897	Napheys, E. C.........	Acetylene gas producing apparatus.
580,650	April 13, 1897	Reynolds, D. J.........	Acetylene gas generator.
581,020	April 20, 1897	Dennis, William H.....	Acetylene gas generating lamp.
581,699	May 4, 1897	Doddridge, A. F........	Apparatus for generating acetylene gas.
582,274	May 11, 1897	Dickerson, Edward N...	Apparatus for gasifying and controlling lique- fied or compressed gas.
582,546	May 11, 1897	Patterson, J. J.........	Apparatus for generating acetylene gas.
582,548	May 11, 1897	Rand, Charles E.......	Process of generating acetylene gas.
583,582	June 1, 1897	Rhind, Frank..........	Gas generating lamp.
583,761	June 1, 1897	Mitchell, F. A..........	Acetylene gas generator.
584,339	June 15, 1897	Exley, J. H............	Apparatus for manufact- uring acetylene gas.
584,772	June 22, 1897	Dickerson, Edward N., and Suckert, J. J.	Apparatus for burning acetylene gas.
584,931	June 22, 1897	Fuller, H. F..........	Acetylene gas generator.
584,946	June 22, 1897	Luckenbach, R.........	Acetylene gas apparatus.
585,625	June 29, 1897	Dougherty, J. F........	Acetylene gas generator.
585,642	June 29, 1897	Gallagher, J. C.........	Gas generator for lamps.
586,194	July 13, 1897	Matthews, C., Jr.......	Acetylene gas apparatus.
586,517	July 13, 1897	Hawley, C. G	Acetylene gas lamp.
587,914	Aug. 10, 1897	Becherel, C. F. J. B....	Apparatus for producing acetylene gas.
588,230	Aug. 17, 1897	Mackusick, E. F.......	Process of generating gas from carbide.
588,535	Aug. 17, 1897	Simonson, F...........	Acetylene gas generator.
588,593	Aug. 24, 1897	Morency, D. C	Apparatus for generating acetylene gas.
589,404	Sept. 7, 1897	Bettini, G.............	Acetylene gas lamp.
589,713	Sept. 7, 1897	Gallagher, J. C........	Process and apparatus for generating acetylene gas.
589,799	Sept. 7, 1897	Taylor, George........	Acetylene gas generator.
590,441	Sept. 21, 1897	Reynolds, D. J.........	Acetylene gas generator.
590,674	Sept. 28, 1897	Strom, A. A...........	Apparatus for generating acetylene gas.
590,592	Sept. 28, 1897	Pyle, H. L., Lichten- stein, L., and Brison, J. C.	Apparatus for generating acetylene gas.
590,941	Sept. 28, 1897	Beck, Charles W	Lamp for generating and burning acetylene gas.
590,955	Oct. 5, 1897	Beck, Charles W.......	Acetylene gas lamp.
591,132	Oct. 5, 1897	Handsby, H. M........	Acetylene gas lamp.
591,367	Oct. 5, 1897	Beck, Charles W.......	Acetylene gas lamp.
592,083	Oct. 19, 1897	Dupee, J. C............	Acetylene gas generator.
592,084	Oct. 19, 1897	Dupee, J. C.......	Acetylene gas generator.

GENERATORS—*Continued.*

NUMBER.	DATE.	INVENTOR.	TITLE.
592,759	Nov. 2, 1897	Bellamy, C. H..........	Apparatus for generating acetylene gas.
593,122	Nov. 2, 1897	Raymond, J. M., and Lemley, L. E.	Acetylene gas generator.
593,628	Nov. 16, 1897	Williams, B. F.........	Generator for acetylene gas.
594,175	Nov. 23, 1897	Hellwig, Otto S........	Acetylene gas generator.
594,826	Nov. 30, 1897	Ferguson, J. S	Acetylene gas generator.
594,849	Dec. 7, 1897	Bettini, G.............	Acetylene gas generator.
595,119	Dec. 7, 1897	Couper, J. H...........	Acetylene gas generator.
595,230	Dec. 7, 1897	Whittemore, L. D., Jr..	Combined lamp and acetylene generator.
595,451	Dec. 14, 1897	Choquette, C. P., and Morin, A. M........	Acetylene gas generator.
595,621	Dec. 14, 1897	Gobron, A............	Acetylene gas generating lamp.
595,668	Dec. 14, 1897	Bryant, H	Acetylene gas machine.
595,816	Dec. 21, 1897	Lebrun, G., and Cornaille, F.	Apparatus for producing acetylene gas.
595,924	Dec. 21, 1897	Ruhe, W. A., and Burbank, H. S.	Acetylene gas apparatus.
596,139	Dec. 28, 1897	Bolton, Werner	Process of generating acetylene gas.
596,144	Dec. 28, 1897	Dolan, E. J..	Burner for rich gases, especially acetylene gas.
596,138	Dec. 28, 1897	Blanchard, D. R.......	Acetylene gas generator.
596,112	Dec. 28, 1897	Henkle, L., and Goddard, F. H............	Acetylene gas burner.
596,577	Jan. 4, 1898	Dolan, E. J.............	Acetylene gas burner.
596,578	Jan. 4, 1898	Dolan, E. J.............	Acetylene gas burner.
596,703	Jan. 4, 1898	Harrison, P. R.........	Acetylene gas generating lamp.
596,937	Jan. 4, 1898	Kerr, J. G.............	Acetylene gas apparatus.
597,291	Jan. 11, 1898	Leede, J.........	Acetylene gas apparatus.
597,495	Jan. 18, 1898	Dolan, E. J........... .	Acetylene gas heater.
597,937	Jan. 25, 1898	Bell, H. J...........	Acetylene generator.
598,048	Jan. 25, 1898	Carter, R. F	Apparatus for producing acetylene gas.
598,213	Feb. 1, 1898	Wilson, C. L.,. Unger, J. W., Muma C., Brosius, A. P., and Kuchel, J. C................	Acetylene gas generator.
598,767	Feb. 8, 1898	Carter, R. F...........	Apparatus for producing acetylene gas.
598,837	Feb. 8, 1898	Appleby, E	Acetylene gas apparatus.
598,868	Feb. 8, 1898	Hardwick, J. L., and Manville, S. O.......	Acetylene gas generator.
599,074	Feb. 15, 1898	Dederick, Z. P.........	Acetylene gas generator.
599,098	Feb. 15, 1898	Hanotier, V., and Hostelet, G.	Apparatus for producing acetylene gas.

GENERATORS—*Continued.*

NUMBER.	DATE.	INVENTOR.	TITLE.
599,198	Feb. 15, 1898	Serres, L.	L a m p for generating acetylene gas.
599,270	Feb. 15, 1898	Stein, A. K	Apparatus for generating acetylene gas.
599,347	Feb. 22, 1898	McMurray, P	Acetylene gas generator.
599,394	Feb. 22, 1898	Laun, H. W. and E. E..	Acetylene gas generator.